Annals of Mathematics Studies

Number 87

Mark G. Krein

SCATTERING THEORY FOR AUTOMORPHIC FUNCTIONS

BY

PETER D. LAX AND
RALPH S. PHILLIPS

PRINCETON UNIVERSITY PRESS

AND

UNIVERSITY OF TOKYO PRESS

PRINCETON, NEW JERSEY
1976

Published in Japan exclusively by
University of Tokyo Press;
In other parts of the world by
Princeton University Press

Printed in the United States of America
by Princeton University Press, Princeton, New Jersey

Library of Congress Cataloging in Publication data will
be found on the last printed page of this book

TABLE OF CONTENTS

PREFACE

Our interest in harmonic analysis of $SL(2,\mathbb{R})$ stems from the fascinating 1972 Faddeev-Pavlov paper [6] in which they showed that the Lax-Phillips theory of scattering could be applied to the automorphic wave equation. After studying [6] we decided to redo this development entirely within the framework of our theory and found new or more straightforward treatments of the following topics:

i) The spectral theory of the Laplace-Beltrami operator over fundamental domains of finite area. It should be noted that the spectral theory as developed in [5] was the starting point for the Faddeev-Pavlov paper.

ii) The meromorphic character over the whole complex plane of the Eisenstein series. In our version this fact is a corollary of a very general abstract theorem in scattering theory.

iii) The Selberg trace formula. The basic ingredients of our derivation are the same as in the original Selberg paper, but the hyperbolic setting is conceptually pleasing and technically a little easier.

Although our monograph contains few new results, we believe that our approach is important enough to justify its publication. Hopefully others will be able to use our methods to obtain new results about subgroups of $SL(2,\mathbb{R})$ or perhaps other groups. Also we hope that our monograph will serve the useful purpose of being an easy introduction for analysts to the algebraic parts of this beautiful field and, at the same time, of providing algebraists and number theorists the necessary analytical tools in a (relatively) painless way. We were encouraged in this by our friend Serge Lang; he and his students, in a seminar on $SL(2,\mathbb{R})$ given in 1973-74, found the analysis used in the more traditional treatments of the spectral theory very hard going. To remedy this problem we made it a practice in the preparation of our monograph, when deciding which analytical details to include and which to leave out, to always ask ourselves the question: "Would Serge want it in?"

We first learned of the work in [6] through lectures and conversations with Ludwig Faddeev during his visit to the United States in the winter of 1972. It is a pleasure to acknowledge our indebtedness to him.

We would like to thank the Energy Research and Development Administration, the Air Force, the National Science Foundation and the John Simon Guggenheim Memorial Foundation for their support and encouragement in carrying out this work.

We dedicate this monograph to Mark G. Krein, one of the mathematical giants of our age, as a tribute to his extraordinarily broad and profound contributions. Like all analysts, we owe him a great deal.

AUGUST, 1975

LIST OF SYMBOLS

$\mathcal{J}$: null-space of A, 105, 109.

$k_{\pm}$: incoming and outgoing translation representers, 15.

$\tilde{k}_{\pm}$: incoming and outgoing spectral representers, 17.

K: bilinear form, 96.

$\mathcal{K}$: orthogonal complement of $\mathcal{D}_- \oplus \mathcal{D}_+$, 26; $\mathcal{K}^a$, 30.

L_o: Laplace-Beltrami operator, 3, 88; L, 94.

$M(\lambda)$: variation in the winding number of $\det \mathcal{S}'(\sigma)$ over $(-\lambda, \lambda)$, 205.

$\mathcal{N}$: auxiliary Hilbert space, 12.

$N(\lambda)$: number of eigenvalues of A of absolute, value $< \lambda$, 205.

$\mathcal{P}$: space spanned by the eigenvectors $\{f_j^{\pm}\}$, 58; $\mathcal{P}_{\pm}$, 61.

$P_{\pm}$: projection operators, 21, 77; $P_{\pm}$, 78; $P''_{\pm}$, 78; $P_{\mathcal{D}}$, 256; $P_{\mathcal{J}}$, 269.

Π: Poincaré plane, 3.

q_j: positive eigenvectors of L, 57; q_o, null vector of L, 65.

Q': projection operator, 59; Q'_-, 68; Q_-, 71; Q''_-, 71.

R: time reversal operator, 69.

$R(t)$: Tiemann matrix of the wave equation, 288; $R_F(t)$, 271.

S: scattering operator, 15; S^a, 30; S', S'', $S_{\pm}$, 66.

$\mathcal{S}$: scattering matrix, 19; $\mathcal{S}^a$, 30; $\mathcal{S}'$, $\mathcal{S}''$, $\mathcal{S}_{\pm}$, 66; $\mathcal{S}_{\pm}^0$, 78.

$T_j\,[\tilde{T}_j]$: translation [spectral] representation, 46.

tr K: trace of K, 220.

$T(r)$: translation operator, 24.

T_Y: truncation operator, 225; T_i^a, 272.

U: unitary group of operators, 12, 59, 109; U_p and U_c, 222; U_o, 235.

Z: associated semigroup of operators, 26, 77; Z^a, 30; Z_-, 67; $Z_{\pm}^0$, 78.

Scattering Theory for Automorphic Functions

§1. INTRODUCTION

The Poincaré plane H, that is the upper half plane:

$$y > 0, \quad -\infty < x < \infty, \quad w = x + iy, \tag{1.1}$$

is a model for non-Euclidean geometry where the role of non-Euclidean motions is taken by the group G of fractional linear transformations

$$w \to \frac{aw + b}{cw + d}, \tag{1.2}$$

$$ad - bc = 1, \quad a, b, c, d \quad \text{real}; \tag{1.3}$$

the matrix $\begin{pmatrix} a & b \\ c & d \end{pmatrix}$ and its negative furnish the same transformation. The Riemannian metric

$$\frac{dx^2 + dy^2}{y^2} \tag{1.4}$$

is invariant under this group of motions. The invariant Dirichlet integral for functions is

$$\iint (u_x^2 + u_y^2)\, dx\, dy; \tag{1.5}$$

and the *Laplace-Beltrami* operator associated with this is

$$L_o = y^2 \Delta = y^2(\partial_x^2 + \partial_y^2). \tag{1.6}$$

The above group of transformations has many so-called *discrete subgroups*; a subgroup Γ is called discrete if the identity is isolated from the other transformations in Γ; these subgroups arise for instance in the conformal theory of Riemann surfaces. An example of a discrete subgroup of interest to number theorists is the *modular group* consisting of transformations with integer a, b, c, d.

A *fundamental domain* F for a discrete subgroup Γ is a subdomain of the Poincaré plane such that every point of Π can be carried into a point of the closure $\overline{F}$ of F by a transformation in Γ and no point of F is carried into another point of F by such a transformation. $\overline{F}$ can be regarded as a manifold where those boundary points which can be mapped into each other by a γ in Γ are identified.

A function f defined on Π is called *automorphic* with respect to a discrete subgroup Γ if

$$f(\gamma w) = f(w) \tag{1.7}$$

for all γ in Γ. Because of (1.7), an automorphic function f is completely determined by its values on $\overline{F}$. If f is continuous, (1.7) imposes a relation between values of f at those pairs of boundary points which can be mapped into each other by a γ in Γ; similarly, if f is once differentiable, (1.7) also imposes a relation between the first derivatives of f at such pairs of boundary points.

The Laplace-Beltrami operator L_0 maps automorphic functions into automorphic functions; the restriction of f to the fundamental domain F has to satisfy the above mentioned boundary relations imposed by (1.7). These relations serve as boundary conditions for the operator L_0; in fact they define L_0 as a self-adjoint operator acting on $L_2(F)$, the space of functions on F square integrable with respect to the invariant measure. The spectral invariants of L_0, that is the point and continuous spectrum and the associated eigenfunctions – proper and generalized – are intimately connected with the subgroup Γ.

Following this line of reasoning, Maas [18] in 1949 showed that the Eisenstein functions were automorphic generalized eigenfunctions for the Laplace-Beltrami operator. A few years later Roelcke [19, 20] used these functions to obtain a spectral representation for this operator in $L_2(F)$ for the modular group; he further observed that the general case could be treated if the analytic continuation for the corresponding Eisenstein functions could be established. This step was accomplished by Selberg who sketched a proof of the analytic continuation in [22]. A more general discussion of Eisenstein series and the spectral decomposition for discrete subgroups can be found in a 1966 paper by Langlands [13]. In the same year Godement [9, 10] published a proof of the spectral theory for the modular group using purely arithmetic methods. In 1967 Faddeev [5] redid the entire spectral theory of L_0 using techniques from perturbation theory, in particular the method of integral equations. More recent presentations of this theory can be found in Elstrodt [4] and Kubota [12].

As far back as 1962, I. M. Gelfand [8] noticed the analogy between the Eisenstein functions and the scattering matrix. However, it was not until 1972 that the connection between harmonic analysis for automorphic functions and scattering theory was firmly established when Faddeev and Pavlov [6] showed that the Lax-Phillips theory of scattering [14, 15] could be directly applied to the wave equation associated with the Laplace-Beltrami operator:[1]

(1.8) $$u_{tt} = y^2\Lambda u + \frac{1}{4}u \ .$$

The Faddeev-Pavlov work relies on the spectral theory for the Laplace-Beltrami operator developed in [5] and is limited to discrete subgroups with fundamental domains of one cusp. Our purpose in this monograph is to develop the complete spectral theory for the Laplace-Beltrami operator, including the analytic continuation of the Eisenstein functions, within the framework of our scattering theory. Specifically we show for any discontinuous group whose fundamental domain is noncompact but has finite

[1] The presence of the term $\frac{1}{4}u$ will be explained in the appendix to this section as well as in Sections 5 and 6.

area that the Laplace-Beltrami operator has continuous spectrum covering $(-\infty, -1/4)$ with a multiplicity equal to the number of cusps of the fundamental domain. We derive a spectral representation for the continuous part of the spectrum in terms of Eisenstein functions, and show that these Eisenstein functions are analytic in a halfplane, meromorphic in the whole plane.

Faddeev and Pavlov studied the relation between the scattering matrix and the Eisenstein functions and, in the case of the modular group, they showed that the poles of the scattering matrix correspond to the zeros of the Riemann zeta function. In the Lax-Phillips theory the poles of the scattering matrix are associated with the exponential decay of certain components of the solutions to the automorphic wave equation. Using this fact, Faddeev and Pavlov reformulated the Riemann hypothesis for the zeta function in terms of the rate of decay of these components of the solutions to the wave equation. Our analysis indicates that this does not constitute a new approach to the problem.

The high point for this circle of ideas is the beautiful trace formula found by Selberg [21] in 1956. Selberg's original treatment is somewhat sketchy and the literature contains several other proofs of this identity. Those recently obtained by T. Kubota [12] and by L. D. Faddeev, V. L. Kalinin and A. B. Venkow [7] apply to all discrete subgroups of $SL(2, \mathbb{R})$ having fundamental domains of finite area. Trace formulas in more general settings have also been constructed by Duflo and Labesse [3] and by Arthur [1].

The organization of this monograph is as follows: In Section 2 we present a self-contained simplified version of our scattering theory which is applicable to the standard wave equation in an exterior domain. However, the automorphic wave equation is not standard, having an indefinite energy form, and this requires substantial modifications in the basic theory; the modified theory is developed in Section 3.

We analyze the Laplace-Beltrami operator on $L_2(F)$ in Section 4. For technical reasons we study, in Section 5, the automorphic wave equation on a function space related to but different from the space $\mathcal{H}$ considered in Section 3. In Section 6 the so-called incoming and outgoing subspaces of data are introduced and their properties are verified. In Section 7 the spectral representation is constructed in terms of the Eisenstein functions and then used to obtain the scattering matrix; the abstract theory is then applied to prove that an Eisenstein function can be continued as a meromorphic function of the parameter in the whole complex plane.

In order to simplify the exposition we have limited the discussion in Sections 4-7 to the modular group. In Section 8 we indicate the changes needed to make the development work for any discrete subgroup of G whose fundamental domain is noncompact and of finite area. Also in this section we obtain a formula for the asymptotic distribution of the point eigenvalues of the Laplace-Beltrami operator.

In Section 9 we obtain the Selberg trace formula from the free space solution to the wave equation (1.8) over the Poincaré plane. This results in a somewhat more direct treatment than the usual proof which is based on invariant integral operators (see [7] and [12]. In the appendix to Section 9 we present another version of the trace formula in which one of the integral terms is replaced by the sum of its residues.

Appendix to Section 1: The wave equation.

In this appendix we derive the Faddeev-Pavlov automorphic wave operator in a way which suggests that the wave equation is a natural tool in the study of harmonic analysis for SL(2, R).

We start with a well-known mapping of SL(2, R) onto the three-dimensional Lorentz group. Given any γ in SL(2, R):

$$\gamma = \begin{pmatrix} a & b \\ c & d \end{pmatrix}, \qquad ad - bc = 1 , \tag{1.9}$$

the mapping $\theta = \theta(\gamma)$:

$$\theta : (\tau', \chi', \eta') \to (\tau, \chi, \eta)$$

is defined by

$$\gamma \begin{pmatrix} \tau' - \chi' & \eta' \\ \eta' & \tau' + \chi' \end{pmatrix} \gamma^t = \begin{pmatrix} \tau - \chi & \eta \\ \eta & \tau + \chi \end{pmatrix} . \tag{1.10}$$

Clearly

$$\theta(\gamma_1 \gamma_2) = \theta(\gamma_1)\theta(\gamma_2) .$$

Taking the determinant on both sides of (1.10) we see by (1.9) that

$$\tau^2 - \chi^2 - \eta^2 = \tau'^2 - \chi'^2 - \eta'^2 ,$$

so that $\theta(\gamma)$ preserves the Lorentz distance. Comparing the trace of both sides of (1.10) we conclude that θ maps the forward light cone C_+:

$$\tau^2 - \chi^2 - \eta^2 > 0 \qquad \text{and} \qquad \tau > 0$$

onto itself. In fact θ is an isomorphism between $SL(2, \mathbf{R})/(\pm 1)$ and that component of the three-dimensional Lorentz group which preserves C_+.

It is easy to determine all transformations γ_o which have i as a fixed point. In fact from

$$\frac{a_o i + b_o}{c_o i + d_o} = i$$

we deduce that this requires

$$a_o = d_o \qquad \text{and} \qquad b_o = -c_o .$$

Using (1.10) we can then easily verify that if i is a fixed point of γ_o, then $(s, 0, 0)$ is a fixed point of $\theta(\gamma_o)$. This observation leads to a mapping Θ of the Poincaré plane Π onto the hyperboloid

$$\tau^2 - \chi^2 - \eta^2 = s^2 . \tag{1.11}$$

For any w in Π choose a γ in $SL(2, \mathbf{R})$ so that

$$\gamma i = w , \tag{1.12}$$

and define

$$\Theta(w) = (\tau, \chi, \eta) = \theta(\gamma)(s,0,0) . \tag{1.13}$$

It follows from the observation made above that $\Theta(w)$ is independent of which γ was used in (1.12). Since $\theta(\gamma)$ is a Lorentz transformation, (1.11) is satisfied by $\Theta(w)$.

Let β be any element in $SL(2,\mathbb{R})$; applying β to (1.12) we get

$$\beta\gamma i = \beta w ;$$

so it follows from (1.13) that

$$\begin{aligned} \Theta(\beta w) &= \theta(\beta\gamma)(s,0,0) \\ &= \theta(\beta)\theta(\gamma)(s,0,0) \\ &= \theta(\beta)\Theta(w) , \end{aligned} \tag{1.14}$$

that is, the mapping Θ intertwines the action of β and $\theta(\beta)$.

Next we transfer the non-Euclidean distance in Π to the hyperboloid (1.11) by defining the distance between $\lambda_1 = \Theta(w_1)$ and $\lambda_2 = \Theta(w_2)$ as the non-Euclidean distance between w_1 and w_2. Since non-Euclidean distance is invariant under β in $SL(2,\mathbb{R})$, it follows from (1.14) that the transferred distance is invariant under Lorentz transformation. But then it must be a multiple of the Lorentz distance. Indeed a brief calculation shows that

$$\frac{dx^2 + dy^2}{y^2} = \frac{d\chi^2 + d\eta^2 - d\tau^2}{s^2} , \tag{1.15}$$

where

$$w = x + iy .$$

It is easy to write down the mapping Θ explicitly; first we solve relation (1.12) for γ; there are many solutions, the simplest one being the one which maps ∞ into ∞, that is for which $c = 0$. For such a γ

(1.12) gives

$$\frac{ai+b}{d} = x + iy \ .$$

so

$$a = dy\,, \qquad b = dx$$

and

$$ad = 1 \ .$$

This set of equations is easily solved:

$$\gamma = \begin{pmatrix} \sqrt{y} & \frac{x}{\sqrt{y}} \\ 0 & \frac{1}{\sqrt{y}} \end{pmatrix} .$$

Substituting this into (1.10) with $\tau' = s$, $\chi' = \eta' = 0$, we get the following expression for the components τ, χ, η of $\Theta(w)$:

(1.16)
$$\tau = \frac{s}{2}\left(\frac{1}{y} + y + \frac{x^2}{y}\right),$$
$$\chi = \frac{s}{2}\left(\frac{1}{y} - y - \frac{x^2}{y}\right),$$
$$\eta = s\,\frac{x}{y} \ .$$

One can regard (1.16) as a mapping of $\Pi \times \mathbf{R}_+$ onto the forward light cone C_+; from (1.15) and (1.16) we deduce

(1.17)
$$ds^2 - \frac{dx^2 + dy^2}{y^2}\, s^2 = d\tau^2 - d\chi^2 - d\eta^2 \ .$$

The change of variables

$$s = e^t$$

turns the left side of (1.17) into

$$e^{2t}\left[dt^2 - \frac{dx^2 + dy^2}{y^2}\right].$$

The Laplace-Beltrami operator of this metric is

$$e^t\partial_t(e^t\partial_t) - e^{2t}y^2(\partial_x^2 + \partial_y^2)$$

while that of the right side of (1.17) is

$$\partial_\tau^2 - \partial_\chi^2 - \partial_\eta^2 .$$

So it follows that if $u(\tau, \chi, \eta)$ satisfies the classical wave equation

$$u_{\tau\tau} = u_{\chi\chi} + u_{\eta\eta} ,$$

then in terms of x, y, t it satisfies

$$(1.18) \qquad e^{-t}\partial_t(e^t\partial_t)u = y^2(u_{xx} + u_{yy}) .$$

Setting

$$v = e^{t/2}u$$

we deduce from (1.18) that v satisfies

$$v_{tt} = y^2(v_{xx} + v_{yy}) + \frac{1}{4}v ;$$

this is the automorphic wave equation (1.8).

§2. AN ABSTRACT SCATTERING THEORY

This section is an essentially self-contained version of the theory presented in the first three chapters of our book [14].

The core of our theory deals with a one-parameter group of unitary operators $U(t)$, $-\infty < t < \infty$, acting on a separable Hilbert space $\mathcal{H}$, in which two closed subspaces $\mathcal{D}_-$ and $\mathcal{D}_+$, called *incoming* and *outgoing spaces* respectively, are distinguished by the following properties:

$$
\text{(2.1)} \qquad
\begin{array}{lll}
\text{i)}_- & U(t)\mathcal{D}_- \subset \mathcal{D}_- & \text{for} \quad t \leq 0 \\
\text{i)}_+ & U(t)\mathcal{D}_+ \subset \mathcal{D}_+ & \text{for} \quad 0 \leq t \\
\text{ii)} & \bigwedge_{t<0} U(t)\mathcal{D}_- = 0 = \bigwedge_{t>0} U(t)\mathcal{D}_+ & \\
\text{iii)} & \overline{\bigvee U(t)\mathcal{D}_-} = \mathcal{H} = \overline{\bigvee U(t)\mathcal{D}_+} \, . &
\end{array}
$$

Here $\wedge$ denotes intersection, $\vee$ union and $\overline{}$ closure.

An example of an outgoing subspace is furnished by the following:

$\mathcal{H} = L_2(R, \mathcal{N})$, the space of all square integrable functions on the real axis, whose values lie in some auxiliary Hilbert space $\mathcal{N}$.

$U(t)$ = translation to the right by t units.

$\mathcal{D}_+ = L_2(R_+, \mathcal{N})$, the subspace of all functions whose support lie on the positive axis.

It is trivial to verify that properties i)$_+$, ii) and iii) are satisfied. Similarly, $D_- = L_2(R_-, \mathcal{N})$ is an example of an incoming subspace within the same framework.

This example furnishes us with the universal model for structures of this kind:

THEOREM 2.1 (Translation Representation Theorem). *Let* $\mathcal{H}$ *be a Hilbert space,* U(t) *a one-parameter group of unitary operators,* $\mathcal{D}_-$ *a subspace of* $\mathcal{H}$ *which has with respect to the group* U *properties* $(2.1)_-$. *Then* $\mathcal{H}$ *can be represented as a space* $L_2(\mathbf{R}, \mathcal{N}_-)$, $\mathcal{N}_-$ *some auxiliary Hilbert space, so that*

i) *The action of* U(t) *in* $\mathcal{H}$ *corresponds to translation to the right by* t *in* $L_2(\mathbf{R}, \mathcal{N}_-)$.

ii) *The correspondence*

$$\mathcal{H} \leftrightarrow L_2(\mathbf{R}, \mathcal{N}_-)$$

is unitary.

iii)$_-$

$$\mathcal{D}_- \leftrightarrow L_2(\mathbf{R}_-, \mathcal{N}_-)$$

Similarly, if $\mathcal{D}_+$ *has properties* $(2.1)_+$, *then* $\mathcal{H}$ *can be represented as* $L_2(\mathbf{R}, \mathcal{N}_+)$ *so that* $\mathcal{D}_+$ *corresponds to* $L_2(\mathbf{R}_+, \mathcal{N}_+)$.

For a proof, see for instance Chapter 2 of [14].

These representations are essentially unique, i.e. unique up to automorphisms of $\mathcal{N}_\pm$. They are called the *incoming*, respectively *outgoing*, *translation representations*. The functions representing a given element of $\mathcal{H}$ are called its *incoming and outgoing representers*.

Fourier transformation turns a translation representation into a spectral representation where the elements of $\mathcal{H}$ are again represented by $\mathcal{N}$-valued functions on $\mathbf{R}$. This shows that the dimension of $\mathcal{N}$ is the multiplicity of the spectrum of U. Since this holds for $\mathcal{N}_-$ as well as $\mathcal{N}_+$, we see that $\dim \mathcal{N}_- = \dim \mathcal{N}_+$; hence $\mathcal{N}_-$ and $\mathcal{N}_+$ are isomorphic and we can identify them, setting

$$\mathcal{N}_- = \mathcal{N}_+ = \mathcal{N} .$$

We remark that in a translation representation *the infinitesimal generator of the group* U *acts as differentiation with respect to the independent variable.*

LEMMA 2.2. *The orthogonal complement of an incoming subspace is outgoing and vice-versa.*

PROOF. Incoming means that conditions $(i)_-$, (ii) and (iii) of (2.1) are satisfied. Since $U(t)$ is unitary,

$$(U(t)d, f) = (d, U(-t)f) ,$$

and it follows from this identity that (a) if $\mathcal{D}_-$ satisfies $(i)_-$, then $\mathcal{D}_-^\perp$ satisfies $(i)_+$, (b) if $\mathcal{D}_-$ satisfies $(ii)_-$, then $\mathcal{D}_-^\perp$ satisfies $(iii)_+$ and (c) if $\mathcal{D}_-$ satisfies $(iii)_-$, then $\mathcal{D}_-^\perp$ satisfies $(ii)_+$. ■

Another way of proving this lemma is to observe that the incoming translation representation for $\mathcal{D}_-$ is also an outgoing translation representation for $\mathcal{D}_-^\perp$.

Later in this monograph we shall have occasion to obtain a translation representation by a direct construction rather than by appealing to the translation representation theorem. This construction proceeds in three states:

1) We devise explicitly a unitary representation for $\mathcal{D}_-$ as $L_2(\mathbf{R}_-, \mathfrak{N})$ where $U(t)$ acts as translation for t negative.

2) We extend the representation to all elements of the form $U(t)d_-$, d_- in $\mathcal{D}_-$, $t > 0$.

3) The representation is extended by continuity to the closure of all elements of the union of $U(t)\mathcal{D}_-$. Since by property $(2.1)_{iii}$ elements of the above form are dense in $\mathcal{H}$, this furnishes the desired representation of $\mathcal{H}$.

If we denote an element of $\mathcal{H}$ by some printed letter, it is convenient to denote its incoming and outgoing representers by the same letter in script with the subscript $-$ and $+$, respectively.

Let k be an element of $\mathcal{H}$, k_- and k_+ its incoming and outgoing representers. The *scattering operator* S *is defined as the operator relating the incoming and outgoing translation representations*:

$$S : k_- \to k_+ . \tag{2.2}$$

The basic properties of the scattering operator are given in the next two theorems:

THEOREM 2.3. *Suppose that properties* i) - iii) *of* (2.1) *hold; then*

(2.3)
i) S *commutes with translation*
ii) S *is a unitary mapping of* $L_2(\mathbf{R}, \mathfrak{N})$ *onto itself.*

PROOF. Property i) follows from property i) of Theorem 2.1, according to which in both incoming and outgoing representations, translation corresponds to the action of U(t). Property ii) follows from the property ii) in Theorem 2.1. ■

We introduce now a further assumption about the subspaces $\mathfrak{D}_-$ and $\mathfrak{D}_+$ which plays a crucial role in what is to follow:

(2.4) iv) $\mathfrak{D}_-$ *and* $\mathfrak{D}_+$ *are orthogonal.*

THEOREM 2.4. *Suppose that in addition to properties* (2.1), *property* (2.4) *holds. Then* $L_2(\mathbf{R}_-)$ *is an invariant subspace of the scattering operator* S, *and* $L_2(\mathbf{R}_+)$ *is invariant for* S^*, *that is*

$$S\, L_2(\mathbf{R}_-) \subset L_2(\mathbf{R}_-) , \tag{2.5$_-$}$$

$$S^* L_2(\mathbf{R}_+) \subset L_2(\mathbf{R}_+) . \tag{2.5$_+$}$$

PROOF. If k_- has its support on $\mathbf{R}_-$, then by property iii)$_-$ of Theorem 2.1, it is the incoming representer of an element k of $\mathfrak{D}_-$. By assumption (2.4), k is orthogonal to $\mathfrak{D}_+$. Since the outgoing representation is unitary, it preserves orthogonality, and so $\mathcal{k}_+$, the outgoing representer of k, is orthogonal to the functions representing $\mathcal{D}_+$. According to part iii)$_+$ of Theorem 2.1, the functions representing $\mathcal{D}_+$ in the outgoing representation fill up $L_2(\mathbf{R}_+)$; therefore $\mathcal{k}_+$, being orthogonal to $L_2(\mathbf{R}_+)$, has its support on $\mathbf{R}_-$, as asserted in (2.5)$_-$. Relation (2.5)$_+$ is a direct consequence of (2.5)$_-$. ∎

Property (2.5) has an interesting intuitive interpretation: Let $\mathcal{k}$ and $\mathcal{h}$ be two functions which are equal on $\mathbf{R}_+$:

$$\mathcal{h}(s) = \mathcal{k}(s) \qquad \text{for} \qquad 0 < s \ .$$

It follows from (2.5)$_-$ applied to $\mathcal{h} - \mathcal{k}$ that then $S\mathcal{k}$ equals $S\mathcal{h}$ on $\mathbf{R}_+$. If we think of the values of functions on $\mathbf{R}_-$ as their *past*, and their values on $\mathbf{R}_+$ as their *future*, then what we have just deduced can be expressed as follows:

The past of $\mathcal{h}$ *has no influence on the future of* $S\mathcal{h}$.

Similarly, we deduce from (2.5)$_+$ that *the future of* $\mathcal{h}$ *has no influence on the past of* $S^*\mathcal{h}$.

This property is called *causality*; an operator with this property is called *causal*.

Any operator S on $L_2(\mathbf{R})$ which commutes with translation can be represented as *convolution* with a distribution which we denote by S(r):

$$(S\mathcal{k})(s) = \int S(r)\mathcal{k}(s-r)\,dr \ . \tag{2.6}$$

S(r) is a distribution whose values are operators mapping $\mathcal{N}$ into $\mathcal{N}$. The property of causality corresponds to S(r) being zero for $r > 0$:

$$S(r) = 0 \qquad \text{for} \qquad r > 0 \ . \tag{2.7}$$

As noted before, a translation representation of the group $U(t)$ is turned into spectral representation by the Fourier transformation. We set

$$\tilde{k}_{\pm}(\sigma) = \frac{1}{\sqrt{2\pi}} \int e^{i\sigma s}\, k_{\pm}(s)\, ds \ . \tag{2.8}$$

THEOREM 2.5 (Spectral Representation Theorem). *Assume that properties* i) - iii) *of* (2.1) *hold. Then*

i) *The correspondences*

$$k \leftrightarrow \tilde{k}_{\pm} \tag{2.9}$$

defined by (2.8) *are unitary mappings of* $\mathcal{H}$ *onto* $L_2(\mathbf{R}, \mathcal{N})$.

ii) *The action of* $U(t)$ *is represented by multiplication by* $e^{i\sigma t}$.

iii) *Denote by* $\mathcal{A}_+$, *respectively* $\mathcal{A}_-$, *the spaces of functions which correspond to* $\mathcal{D}_+$ *and* $\mathcal{D}_-$ *under the mapping* $(2.8)_+$ *and* $(2.9)_-$ *respectively. These spaces can be characterized as follows:*

$\mathcal{A}_+[\mathcal{A}_-]$ *consists of the boundary values on the real axis of* $\mathcal{N}$-*valued functions which are holomorphic in the upper* [*lower*] *half-plane* $\operatorname{Im}\sigma > 0$ $[\operatorname{Im}\sigma < 0]$, *and whose square integral along the lines* $\operatorname{Im}\sigma = \text{const}$ *is a decreasing function of* $|\operatorname{Im}\sigma|$.

PROOF. These results are mere corollaries of the corresponding assertions i), ii) and iii) of the Translation Representation Theorem 2.1. To deduce i) we appeal to Plancherel's observation that Fourier transformation is unitary on $L_2(\mathbf{R}, \mathcal{N})$. To deduce ii) we recall that under Fourier transformation translation becomes multiplication by $e^{i\sigma t}$. To deduce iii) we have to recall that Fourier transformation carries $L_2(\mathbf{R}_+)$ and $L_2(\mathbf{R}_-)$ onto the space of analytic functions $\mathcal{A}_+$ and $\mathcal{A}_-$, respectively. That the Fourier transform of a function in $L_2(\mathbf{R}_-)$ belongs to $\mathcal{A}_-$ can be directly verified from formula (2.8) for the Fourier transform; that the Fourier inverse of $\mathcal{A}_-$ belongs to $L_2(\mathbf{R}_-)$ is the Paley-Wiener theorem. ∎

The representations (2.9) are called the *incoming and outgoing spectral representations*, respectively. The functions $\tilde{k}_{\pm}(\sigma)$ are called the *incoming and outgoing spectral representers of* k.

We remark that in these spectral representations the infinitesimal generator of the group U acts as multiplication by $i\sigma$.

As noted above, in the spectral representations the Hilbert space $\mathcal{H}$ is represented as $L_2(\mathbf{R}, \mathcal{N})$ and U(t) as multiplication by $e^{i\sigma t}$. This shows that *the spectrum of* U(t) *is absolutely continuous, with uniform multiplicity over the whole real axis equal to the dimension of* $\mathcal{N}$.

Next we turn to investigating the action of the scattering operator S in the spectral representation, under the additional assumption that $\mathcal{D}_+$ and $\mathcal{D}_-$ are orthogonal. We denote by $\mathcal{S}$ the operator

$$\mathcal{S} : \tilde{k}_- \to \tilde{k}_+ . \tag{2.10}$$

Clearly, since translation representation is related to spectral representation by the Fourier transform F,

$$\mathcal{S} = FSF^{-1} . \tag{2.11}$$

The basic properties of $\mathcal{S}$ are expressed in the following theorem which parallels the results of Theorems 2.3 and 2.4:

THEOREM 2.6. *If properties* i) - iv) *of* (2.1) *and* (2.4) *hold then*

i) $\mathcal{S}$ *is multiplication by a function* $\mathcal{S}(\sigma)$ *whose values are operators mapping* $\mathcal{N}$ *into* $\mathcal{N}$.

(2.12) ii) *The values of* $\mathcal{S}(\sigma)$ *are unitary operators.*

iii) $\mathcal{S}(\sigma)$ *is the boundary value of an analytic function defined in the lower half plane whose values are contraction operators mapping* $\mathcal{N}$ *into* $\mathcal{N}$.

REMARK. The function $\mathcal{S}(\sigma)$ is called the *scattering matrix*.

There are several ways of going about the proof of this theorem. One is based on representation (2.6) of S as a convolution operator, from which one concludes that $\mathcal{S}$ acts as multiplication by the Fourier transform of S:

$$\mathcal{S}(\sigma) = \int S(s)\, e^{i\sigma s}\, ds \ . \tag{2.13}$$

This proves part i) of the theorem.

Formula (2.11) shows that $\mathcal{S}$ is the product of three unitary operators. Therefore $\mathcal{S}$ itself is unitary, which together with i), implies (modulo some real variables theory) part ii) of the theorem.

According to (2.7), causality implies that S(s) is supported on the negative axis. It follows then that the Fourier transform of S(s) is analytic in the lower half plane, as asserted in the first half of part iii). ∎

The derivation we have just given of part i) of the theorem and the analyticity of $\mathcal{S}(z)$ is somewhat formal. There are several ways of making it rigorous and quantitative; here we follow a method described previously in [16]:

The starting point of that method is the observation that the Fourier transform of a function S(r) can be obtained by convolving it with an exponential function. Let z be a complex number, n an element of $\mathfrak{N}$; then

$$S(e^{-izs}n) = S * e^{-izs}n = \int S(r)\, e^{-iz(s-r)}n\, dr \tag{2.14}$$

$$= e^{-izs} \int S(r)\, e^{izr} n\, dr = e^{-izs}\, \mathcal{S}(z)\, n \ .$$

As it stands, (2.14) isn't meaningful because the scattering operator as so far defined acts only on square integrable functions, and the exponential function is not square integrable for any value of z. This difficulty is easily overcome by extending the domain of the operator S as follows:

Let $k(s)$ be any $\mathfrak{N}$-valued function on $\mathbf{R}$ which is L_2 on $\mathbf{R}_+$ and locally L_2 on $\mathbf{R}_-$; we denote this space of functions as follows:

$$L_2^{loc}(\mathbf{R}_-) \oplus L_2(\mathbf{R}_+) \equiv L_2^{-loc}(\mathbf{R}) . \tag{2.15}$$

This is a metric space under the seminorms

$$\|k\|_j = \left(\int_{-j}^{\infty} \|k(s)\|_{\mathfrak{N}}^2 \, ds\right)^{\frac{1}{2}}, \qquad j = 1, 2, \cdots . \tag{2.16}$$

Clearly, the function space (2.15) is the completion of $L_2(\mathbf{R})$ with respect to the metric induced by the semi-norms (2.16).

Next we show that if S is causal, then S is contractive with respect to each of the seminorms (2.16). To see this, take any function k in $L_2(\mathbf{R})$; we denote its truncation below $-j$ by k_j:

$$k_j(s) = \begin{cases} k(s) & \text{for} \quad -j < s \\ 0 & \text{for} \quad s < -j \end{cases} .$$

By causality Sk and Sk_j agree for $s > -j$; so

$$\|Sk\|_j = \|Sk_j\|_j \le \|Sk_j\| = \|k_j\| = \|k\|_j$$

where we have used the isometry of S with respect to the L_2 norm, and the fact that $\|k_j\| = \|k\|_j$. This shows that S as defined over $L_2(\mathbf{R})$ is continuous with respect to the metric induced by the seminorm (2.16); therefore S can be extended by continuity as an operator mapping the function space (2.15) into itself. S thus extended retains – obviously – the property of commuting with translation. It is equally obvious that S extended satisfies the following form of causality:

$$SL_2^{loc}(\mathbf{R}_-) \subset L_2^{loc}(\mathbf{R}_-) \, . \tag{2.17}$$

The incoming translation representation establishes a one-to-one correspondence

$$\mathcal{H} \leftrightarrow L_2(\mathbf{R}) \, .$$

We use this correspondence to transfer from $L_2(\mathbf{R})$ to $\mathcal{H}$ the seminorms (2.16) and the metric induced by them; *we denote by* $\overline{\mathcal{H}}$ *the completion of* $\mathcal{H}$ *in this metric*. Clearly, the above correspondence can be extended by continuity to

$$\overline{\mathcal{H}} \leftrightarrow L_2^{-loc}(\mathbf{R}) \, .$$

This is the *incoming translation representation for* $\overline{\mathcal{H}}$.

Having already shown that the scattering operator S can be extended to $L_2^{-loc}(\mathbf{R})$, we *define* the *outgoing translation representation of an element in* $\overline{\mathcal{H}}$ by applying S to its incoming translation representation.

Most of the interesting operators on $\mathcal{H}$ can be extended to $\overline{\mathcal{H}}$; we shall continue to denote the extended operator by the same symbol as the original one.

Translation to the right by t is a continuous operator in the function space $L_2^{-loc}(\mathbf{R})$. Therefore it can be transferred to $\overline{\mathcal{H}}$ via the incoming translation representation, and constitutes an extension to $\overline{\mathcal{H}}$ of the operators U(t). The extended operators form a one parameter group; the generator A of this group, is the strong limit of $(U(h)-I)/h$. The domain of A consists of those elements of $\overline{\mathcal{H}}$ for which the limit exists.

A pair of useful operators in $\mathcal{H}$ are the projections P_+ and P_- which remove the $\mathcal{D}_+$, respectively $\mathcal{D}_-$, component of the element on which they act. In the outgoing, respectively incoming, translation representation these operators act as truncation above, respectively below, zero. Clearly, these operators are contractions with respect to each seminorm (2.16) and so they can be extended by continuity to $\overline{\mathcal{H}}$.

For an element f of $\overline{\mathcal{H}}$ we shall denote by f_+, respectively f_-, its $\mathcal{D}_+$, respectively $\mathcal{D}_-$ component; i.e.

$$f_+ = f - P_+ f \, , \quad f_- = f - P_- f \, .$$

The following result is useful:

Since S is continuous on $L_2(\mathbf{R})$,

$$(S\mathscr{k}, \ell) = (\mathscr{k}, S^*\ell) \tag{2.18}$$

for all ℓ and $\mathscr{k}$ in $L_2(\mathbf{R})$.

CLAIM. (2.18) *is true for all* $\mathscr{k}$ *in* $L_2^{-\text{loc}}(\mathbf{R})$ *and all* ℓ *in* $L_2(\mathbf{R}_+)$.

PROOF. It follows from $(2.5)_+$ that $S^*\ell$ belongs to $L_2(\mathbf{R}_+)$. This implies that the right side of (2.18) depends continuously on $\mathscr{k}$ in all the seminorms. We deduce from $(2.5)_-$ that the left side of (2.18) does not depend on values of $\mathscr{k}$ on $\mathbf{R}_-$; this shows that also the left side of (2.18) depends continuously on $\mathscr{k}$ in all the seminorms. But then the validity of (2.18) for all $\mathscr{k}$ in $L_2^{-\text{loc}}(\mathbf{R})$ follows by continuity. ∎

Next we describe a large class of eigenelements of A in $\overline{\mathcal{H}}$. We start with the observation that for $\operatorname{Im} z < 0$, and any n in $\mathfrak{N}$ the exponential functions

$$e^{-izs}n \,, \tag{2.19$_-$}$$

belong to the function space $L_2^{-\text{loc}}(\mathbf{R})$, and are eigenfunctions of translation by the amount t, with eigenvalue e^{izt}. Denote by $e = e(z) = e(z,n)$ the element of $\overline{\mathcal{H}}$ which the function $(2.19)_-$ represents in the incoming translation representation; it follows that e(z) is an eigenfunction of U(t):

$$U(t)\,e(z) = e^{izt}\,e(z) \,.$$

Consequently the outgoing translation representer $e_+(s)$ of $e(z,n)$ is an eigenfunction of translation, with eigenvalue e^{izt}:

$$e_+(s-t) = e^{izt}e_+(s) \,.$$

It follows from this that e_+ is of the form

$$e^{-izs}m \,, \tag{2.19$_+$}$$

m some element of $\mathcal{N}$; m is a function of n and z. Clearly m is a linear function of n, so we can write it in the form

(2.20) $$m = \mathcal{S}(z)n .$$

In identifying the operator linking n to m with the scattering matrix $\mathcal{S}$ we are guided by the intuitive result (2.14).

Relations (2.19) and (2.20) are the basis of our study of the analytic properties of $\mathcal{S}(z)$.

We start by showing that $\mathcal{S}(z)$ is analytic in the lower half plane. In (2.18) set

(2.21) $$k(s) = e^{-izs}n$$
$$\ell(s) = \begin{cases} 0 & \text{for} \quad s < 0 \\ e^{\bar{\tau}s}p & \text{for} \quad 0 < s \end{cases}$$

where $\operatorname{Re}\tau < 0$ and p is some element of $\mathcal{N}$. We recall from $(2.19)_+$ and (2.20) that $Sk = e^{-izs}\mathcal{S}(z)n$. We denote $S^*\ell$ by h, and recall from $(2.5)_+$ that h is supported on $\mathbf{R}_+$. Setting all this information into (2.18), that is

$$(Sk, \ell) = (k, S^*\ell) ,$$

we get

$$\int_0^\infty (\mathcal{S}(z)n, p)_{\mathcal{N}}\, e^{(-iz+\tau)s}\, ds = \int_0^\infty e^{-izs}(n, h(s))_{\mathcal{N}}\, ds .$$

From this we deduce that

(2.22) $$(\mathcal{S}(z)n, p)_{\mathcal{N}} = (iz-\tau)\int_0^\infty e^{-izs}(n, h(s))_{\mathcal{N}}\, ds .$$

Clearly the right side is an analytic function of z in $\operatorname{Im} z < 0$. This proves that $\mathcal{S}(z)$ is weakly – and therefore strongly – analytic in the lower half plane.

Formula (2.22) serves to estimate $\mathcal{S}(z)$; applying the Schwarz inequality we get

$$|(e^{-izs}n, k)| \leq \left\{ \int_0^\infty \|e^{-izs}n\|_{\mathcal{N}}^2 \, ds \right\}^{\frac{1}{2}} \|k\| = \frac{\|n\|_{\mathcal{N}}}{\sqrt{2|\mathrm{Im}\, z|}} \|k\| .$$

Since S is unitary, $\|k\| = \|\ell\|$; from definition (2.21) of ℓ we get

$$\|\ell\| = \frac{1}{\sqrt{2|\mathrm{Re}\, \tau|}} \|p\|_{\mathcal{N}} .$$

So altogether we can estimate the right side of (2.22) as follows

$$|(\mathcal{S}(z)n, p)_{\mathcal{N}}| \leq \frac{|iz - \tau|}{2\sqrt{\mathrm{Im}\, z\, \mathrm{Re}\tau}} \|n\|_{\mathcal{N}} \|p\|_{\mathcal{N}} .$$

Setting $\tau = i\bar{z}$ we get $\mathrm{Re}\tau = \mathrm{Im}\, z$ and hence

$$|(\mathcal{S}(z)n, p)_{\mathcal{N}}| \leq \|n\|_{\mathcal{N}} \|p\|_{\mathcal{N}} .$$

Since this holds for all pairs of vectors n and p in $\mathcal{N}$, we deduce that $\mathcal{S}(z)$ is a contraction, as asserted in part iii) of Theorem 2.6.

Finally we wish to connect definition (2.20) of the scattering matrix with our earlier definition (2.11). To do so we take any function k_- in $L_2(\mathbf{R})$ whose support lies in $s < a$, a any number. By causality the support of $k_+ = Sk_-$ also lies in $s < a$. Since S commutes with translation, we have

$$ST(r)k_- = T(r)k_+ , \tag{2.23}$$

where $T(r)$ is translation to the right by r units.

Let z be any complex number with $\mathrm{Im}\, z < 0$; multiply (2.23) by e^{-izr} and integrate with respect to r from $-R$ to R. Since $T(r)$ is strongly continuous in r, we can carry out the integration on the left inside S, and we get

$$S \int_{-R}^{R} k_{-}(s-r)\, e^{-izr}\, dr = \int_{-R}^{R} k_{+}(s-r)\, e^{-izr}\, dr .$$

Introducing $s-r = t$ as new variable of integration we get

$$S e^{-izs} \int_{s-R}^{s+R} k_{-}(t)\, e^{izt}\, dt = e^{-izs} \int_{s-R}^{s+R} k_{+}(t)\, e^{izt}\, dt .$$

Let $R \to \infty$; since both $k_{-}(t)$ and $k_{+}(t)$ are zero for $t > a$, the functions $e^{izt} k_{\pm}(t)$ are L_1 integrable and their integrals over $(s-R, s+R)$ converge boundedly to $\tilde{k}_{\pm}(z)$. Because of the exponential factor e^{-izs}, the function on the right tends to $e^{-izs} \tilde{k}_{+}(z)$ in each of the seminorms defining the topology of $L_2^{-loc}(\mathbf{R})$ and the function inside S on the left similarly tends to $e^{-izs} \tilde{k}_{-}(z)$. Since S is continuous in this topology, we conclude that

$$S e^{-izs} \tilde{k}_{-}(z) = e^{-izs} \tilde{k}_{+}(z) .$$

Recalling definition $(2.19)_{\pm}$, (2.20) of $\mathcal{S}(z)$ with $n = \tilde{k}_{-}(z)$, $m = \tilde{k}_{+}(z)$, we see that

$$\mathcal{S}(z) \tilde{k}_{-}(z) = \tilde{k}_{+}(z) , \tag{2.24}$$

i.e. that $\mathcal{S}(z)$ as defined by (2.20) acts multiplicatively on the incoming spectral representation to produce the outgoing spectral representation. This has been established for all elements whose incoming translation representation is zero for $s > a$. Since these elements are dense in $\mathcal{H}$, one can deduce easily that (2.24) holds for all k in $\mathcal{H}$, for almost all real z. This completes our discussion of Theorem 2.6. ■

We recall now the projection operators P_{+} and P_{-} defined earlier as the operators which remove the $\mathcal{D}_{+}$ and $\mathcal{D}_{-}$ components respectively. In terms of these projections we define the following one-parameter family of operators $Z(t)$:

(2.25) $$Z(t) = P_+ U(t) P_- , \qquad t \geq 0 .$$

These operators play a central role in the whole theory; they have the following properties:

THEOREM 2.7. *Suppose properties* (2.1) *and* (2.4) *are satisfied. Then*

a) *The operators* $Z(t)$, $t \geq 0$, *form a strongly continuous semigroup of contraction operators on*

(2.26) $$\mathcal{K} = \mathcal{H} \ominus (\mathcal{D}_- \oplus \mathcal{D}_+)$$

b) $\lim_{t \to \infty} Z(t)k = 0$ *for every* k *in* $\mathcal{H}$.

c) $Z(t)$ *annihilates both* $\mathcal{D}_-$ *and* $\mathcal{D}_+$.

PROOF. First we show that $Z(t)$ maps $\mathcal{H}$ into $\mathcal{K}$. Clearly, the range of $Z(t)$ is orthogonal to $\mathcal{D}_+$, for the leftmost factor of the product (2.25) defining Z is P_+, whose range is orthogonal to $\mathcal{D}_+$. To show orthogonality to $\mathcal{D}_-$ we observe that since $U(t)$ forms a one parameter group, $U(-t) = U(t)^{-1}$ and that since $U(t)$ is unitary,

$$(U(t)f, g) = (f, U(-t)g) .$$

In particular $(U(t)f, \mathcal{D}_-) = (f, U(-t)\mathcal{D}_-)$ and hence if f is orthogonal to $\mathcal{D}_-$, property i)$_-$ of (2.1) implies that for $t > 0$, $U(t)f$ is also orthogonal to $\mathcal{D}_-$.

Now consider the action of the $Z(t)$ defined as the triple product (2.25): The first factor P_- projects any element of $\mathcal{H}$ into the orthogonal complement of $\mathcal{D}_-$, the second factor $U(t)$ keeps it there, and so does the third factor P_+, since the $\mathcal{D}_+$ component, which is removed by P_+, is by (2.4) orthogonal to $\mathcal{D}_-$. This completes the proof that the range of Z lies in $\mathcal{K}$.

To prove the semigroup property we shall use property i)$_+$ of (2.1), according to which for $t > 0$, $U(t)$ maps $\mathcal{D}_+$ into itself. We can write this fact as

(2.27) $$P_+ U(t)(I-P_+) = 0 .$$

Consider now

$$Z(t)Z(s) = P_+ U(t) P_- P_+ U(s) P_- ;$$

Since we have already shown that $Z(s)$ maps into $\mathcal{K}$, we can omit the factor P_- in the middle, and write

$$\begin{aligned} Z(t)Z(s) &= P_+ U(t) P_+ U(s) P_- \\ &= P_+ U(t) U(s) P_- - P_+ U(t)[I-P_+] U(s) P_- . \end{aligned}$$

Using (2.27) we see that the second term is 0; and hence from the group property of U, we obtain

$$Z(t)Z(s) = P_+ U(t) U(s) P_- = P_+ U(t+s) P_- = Z(t+s) ;$$

this is the semigroup property of Z.

The three factors appearing in the product (2.25), two projections and one unitary operator, are all contractions. Therefore so is their product; this completes part a).

To prove part b) we make use of part iii) of (2.1), according to which elements of the form

(2.28) $$f = U(s)g, \qquad g \text{ in } \mathcal{D}_+ ,$$

are dense in $\mathcal{H}$. For an element of this form we have by property $\text{i})_+$ of (2.1) that for $t+s > 0$

$$U(t)f = U(t)U(s)g = U(t+s)g$$

belongs to $\mathcal{D}_+$. Therefore for $t+s > 0$,

$$P_+ U(t) f = 0 .$$

This shows that

(2.29) $$\lim P_+ U(t) f = 0$$

for all elements of form (2.28). Since these elements are dense in $\mathcal{H}$, and since the operator $P_+U(t)$ is a contraction, it follows that (2.29) holds for all f in $\mathcal{H}$. Property b):

$$\lim_{t\to\infty} Z(t)k = 0$$

follows from this by setting $f = P_-k$.

Property c), the annihilation of $\mathcal{D}_+$ and $\mathcal{D}_-$, is obvious.

The effect of $Z(t)$ on $\mathcal{K}$ is particularly simple in the outgoing translation representation. On $\mathcal{K}$ the operator P_- acts as the identity, so

$$Z(t) = P_+U(t) \quad \text{on} \quad \mathcal{K}\,.$$

In the outgoing translation representation $U(t)$ acts as right shift, and P_+ chops off that part of the function which is supported on $\mathbf{R}_+$. The semigroup property follows immediately from this representation of Z. This completes our discussion of Theorem 2.7. ∎

Denote by B the infinitesimal generator of the semigroup Z:

$$Bf = \lim_{t\to 0} \frac{Z(t)f-f}{t}\,; \tag{2.30}$$

the domain of B consists of all f in $\mathcal{K}$ for which the limit exists.

The resolvent of B can be expressed as the Laplace transform of Z:

$$(\lambda - B)^{-1} = \int_0^\infty Z(t)e^{-\lambda t}\,dt\,, \tag{2.31}$$

valid for all λ for which the integral on the right side converges in the strong topology. In particular, for a contraction semigroup such as Z, (2.31) hold for all λ with $\mathrm{Re}\lambda > 0$. *So the open right half plane belongs to the resolvent set of* B.

It follows that *the spectrum of* B *is confined to the left half plane*:

$$\operatorname{Re} \sigma(B) \leq 0 .$$

In fact, a little more is true:

CLAIM: *The imaginary axis is free of the point spectrum of* B.

PROOF. If μ is in the point spectrum of B, and k the eigenvector:

$$Bk = \mu k ,$$

then k is also an eigenvector of Z(t):

$$Z(t)k = e^{\mu t} k .$$

If μ were purely imaginary, we would have

$$\|Z(t)k\| = \|k\| ,$$

contrary to part b) of Theorem 2.7. ■

We recall that we denoted the infinitesimal generator of the group U(t) by A. Substituting the definition (2.25) of Z into (2.31) we deduce the following relation between the infinitesimal generators:

$$(2.32) \qquad (\lambda - B)^{-1} = P_+(\lambda - A)^{-1} P_- \quad \text{for} \quad \operatorname{Re}\lambda > 0 .$$

Another relation can be deduced as follows:

Suppose f is orthogonal to $\mathfrak{D}_-$; then P_+f belongs to $\mathcal{K}$, $P_-P_+f = P_+f$ and by (2.27)

$$Z(t)P_+f = P_+U(t)P_+f = P_+U(t)f .$$

Suppose in addition that f belongs to the domain of A. Differentiating the above relation at $t = 0$ we get

$$(2.33) \qquad BP_+f = P_+Af ,$$

valid for all f in the domain of A which are orthogonal to $\mathfrak{D}_-$.

The following minor modification of the setup just described is very useful in a technical way:

Let a be any positive real number and denote by $\mathcal{D}_+^a$ and $\mathcal{D}_-^a$ the following translates of $\mathcal{D}_\pm$:

(2.34) $$\mathcal{D}_+^a = U(a)\mathcal{D}_+ , \quad \mathcal{D}_-^a = U(-a)\mathcal{D}_- .$$

It is obvious that these smaller subspaces satisfy properties (2.1) postulated for outgoing and incoming subspaces, and that they are orthogonal. We may therefore define analogous operators S^a and Z^a. The change in the scattering operator results from the shift in the origin of $-a$ in the incoming translation representation and of $+a$ in the outgoing translation representation. As a consequence

(2.35) $$\mathcal{S}^a(\sigma) = e^{-2ia\sigma}\mathcal{S}(\sigma) .$$

On the other hand

(2.36) $$Z^a(t) = P_+^a U(t) P_-^a ,$$

where $P_\pm^a$ are projections which remove the $\mathcal{D}_+^a$ and $\mathcal{D}_-^a$ components. Z^a forms a semigroup of operators on the subspace

$$\mathcal{K}^a = \mathcal{H} \ominus (\mathcal{D}_-^a \oplus \mathcal{D}_+^a) .$$

The subspace $\mathcal{K}^a$ is slightly larger than $\mathcal{K}$; this difference gives us extra room for maneuvering.

Using the relation (2.32), we deduce that the infinitesimal generators B^a of Z^a are related to each other as follows: For $a < b$ and for $\operatorname{Re}\lambda > 0$,

$$(\lambda - B^a)^{-1} = P_+^{a,b}(\lambda - B^b)^{-1} P_-^{a,b} ,$$

where $P_\pm^{a,b}$ denote the projection onto $\mathcal{D}_\pm^a \ominus \mathcal{D}_\pm^b$. It follows from this formula that if the resolvent of B^b can be continued from the right half plane into some open set ρ_0, then so can the resolvent of B^a for $a < b$.

Actually, for the situation at hand (as well as in applications to mathematical physics), the resolvent set of B^a is independent of the parameter a.

Both Z and the scattering operator S were constructed out of the same ingredients U(t), $\mathfrak{D}_+$ and $\mathfrak{D}_-$; it is therefore to be expected that there is an intimate relation between the two. The next theorem exhibits such a relation:

THEOREM 2.8. *Denote by* B^a *the infinitesimal generator of* Z^a, a *any positive number, and denote by* $\rho_0(B^a)$ *the principal component of the resolvent set of* B^a, *that is that component of the resolvent set which contains the right half plane.*

ASSERTION. *The scattering matrix* $\mathcal{S}(z)$ *can be continued analytically from the lower halfplane into* $-i\rho_0(B^a)$.

PROOF. We shall use the eigenfunctions e(z, n) of A in $\overline{\mathcal{H}}$ constructed earlier. We recall that z is any complex number in the lower half plane, n any element of $\mathfrak{N}$, and that the incoming and outgoing translation representations of e(z, n), given by $(2.19)_{\pm}$, (2.20), are as follows:

(2.37) $$e_- = e^{-izs}\, n\,, \quad e_+ = e^{-izs}\, m\,, \quad m = \mathcal{S}(z)\, n\ .$$

Now let $\xi(s)$ be a scalar cutoff function of the following sort:

(2.38) $$\xi(s) = \begin{cases} 0 & \text{for} \quad s < -a \\ 1 & \text{for} \quad 0 < s \\ \text{smooth in between}\ . \end{cases}$$

We define the function $f_-(s)$ by

(2.39) $$f_-(s) = \xi(s)\, e_-(s)\ .$$

Because of the cutoff, f_- belongs to $L_2(\mathbf{R}, \mathfrak{N})$ and therefore is the incoming translation representer of some element f of $\mathcal{H}$.

We claim that the cutoff doesn't effect the $\mathcal{D}_+$ component of the outgoing representer, i.e. that the representers of e and f agree on $\mathbf{R}_+$. For it follows from (2.38), (2.39) that the difference $e_- - f_-$ has its support on $\mathbf{R}_-$; and it follows then from the causality property (2.17) that also $e_+ - f_+$ is supported on $\mathbf{R}_-$, which shows that $e_+(s) = f_+(s)$ on $\mathbf{R}_+$.

Now $e_+(s)$ is given by (2.37) so that

$$f_+(s) = e^{-izs}\mathcal{S}(z)n \qquad \text{for} \qquad 0 < s \,. \tag{2.40}$$

Since $\xi(s)$ is zero for $s < -a$, so is $f_-(s)$; this makes f orthogonal to $\mathcal{D}^a_-$. Moreover $\xi(s)$ was chosen to be differentiable so that f_- belongs to the domain of ∂_s:

$$-\partial_s f_- = \xi iz e_- - \xi' e_- = iz f_- - k_- \,, \tag{2.41}$$

where

$$k_- = \xi' e_- \qquad \text{and} \qquad \xi' = \frac{\partial \xi}{\partial s} \,. \tag{2.42}$$

Since $\xi' = 0$ for $s < -a$ and $s > 0$, it follows that k belongs to $\mathcal{K}^a$. Since in a translation representation $-\partial_s$ represents the action of A, we conclude that f belongs to the domain of A, and that

$$Af = izf - k \,,$$

k being that element of $\mathcal{K}^a$ which is represented by (2.42) in the incoming representation. We apply P^a_+ to both sides above and get

$$P^a_+ Af = izP^a_+ f - k \,; \tag{2.43}$$

here we have used the fact that since k belongs to $\mathcal{K}^a$, $P^a_+ k = k$.

We have already shown that f is orthogonal to $\mathcal{D}^a_-$, and that f belongs to the domain of A. Thus we can apply relation (2.33), which asserts that

$$B^a P_+^a f = P_+^a A f .$$

Substituting this into (2.43) we get

$$B^a P_+^a f = izP_+^a f - k .$$

We introduce the notation

(2.44) $$P_+^a f = g ;$$

then the above relation can be rewritten as

$$B^a g = izg - k .$$

Since iz lies in the resolvent set of B^a, this equation can be used to express g in terms of k:

(2.45) $$g = (iz - B^a)^{-1} k .$$

The projection P_+^a removes the $\mathcal{D}_+^a$ component; its action in the outgoing translation representation is truncation above $s = a$. Since $g = P_+^a f$, it follows from (2.40) that

(2.46) $$g_+(s) = e^{-izs} \mathcal{S}(z) n \qquad \text{for} \qquad 0 < s < a .$$

Define the element w of $\mathcal{H}$ by

(2.47) $$w_+(s) = \begin{cases} e^{-i\bar{z}s} p & \text{for} \quad 0 < s < a \\ 0 & \text{elsewhere} , \end{cases}$$

p any element of $\mathcal{N}$. From (2.46) and (2.47) we get that

$$(g, w) = (\mathcal{S}(z) n, p)_{\mathcal{N}} a .$$

Using formula (2.45) for g we obtain

(2.48) $$(\mathcal{S}(z) n, p)_{\mathcal{N}} = \frac{1}{a} ((iz - B^a)^{-1} k, w) .$$

Formula (2.42) shows that k depends analytically on z and formula (2.47) shows that w depends analytically on $\bar{z}$; so it follows from (2.48) that $\mathcal{S}(z)$ can be continued weakly analytically into $-i\rho_0$ if the resolvent of B can be continued analytically into ρ_0. This completes the proof of Theorem 2.8. ■

Formula (2.48) can be used to estimate the norm of $\mathcal{S}(z)$. We deduce first from (2.42) that

$$\|k\|^2 = \int_{-a}^{0} (\xi')^2 e^{2(\mathrm{Im}\, z)s} \, ds \, \|n\|_{\mathcal{N}}^2 .$$

Choosing ξ optimally we get

$$\|k\|^2 = \frac{2 \,\mathrm{Im}\, z}{\exp(2a \,\mathrm{Im}\, z) - 1} \|n\|_{\mathcal{N}}^2 .$$

Obviously

$$\|w\|^2 = \frac{1 - \exp(-2a \,\mathrm{Im}\, z)}{2 \,\mathrm{Im}\, z} \|p\|_{\mathcal{N}}^2 .$$

Estimating the right side of (2.48) by the Schwarz inequality therefore gives

(2.49) $$\|\mathcal{S}(z)\|_{\mathcal{N}} \le \frac{1}{a} e^{-a \,\mathrm{Im}\, z} \|(iz - B^a)^{-1}\| .$$

If the resolvent of B^a is meromorphic in the whole plane, then from Theorem 2.8 and the estimate (2.49) we conclude that the same is true for $\mathcal{S}$:

THEOREM 2.9. *If the resolvent of* B^a *is meromorphic in the whole complex plane, then* $\mathcal{S}(z)$ *can be continued as a meromorphic function into the whole plane.*

For the automorphic wave equation it will be shown that the resolvent of B^a is a compact operator from which it follows that the resolvent of B^a is meromorphic in the whole plane.

REMARK. It is not difficult to show (see Chapter 3 of [14] or better Theorem 5.5 of [16]) that *the poles of* $\mathcal{S}(z)$ *coincide with* $-i\sigma(B^a)$ if the resolvent of B^a is meromorphic in the whole plane. Since this result does not play a significant role in this monograph we omit its proof. However, it follows from this and the relation (2.35) that the spectrum of B^a does not depend on the parameter a.

We introduced the generalized eigenvectors $e(z,n)$ in this section as a means of studying the analytic properties of the scattering matrix. It turns out, both in applications to mathematical physics and in the present situation, that these generalized eigenvectors which so far have been only abstract entities in the space $\bar{\mathcal{H}}$, can be realized as concrete pointwise defined generalized eigenfunctions of the generator A; this realization leads to an explicit formula for the scattering matrix. In Section 7 we treat this matter in detail for the automorphic wave equation; the following theorem will be of use there:

THEOREM 2.10. *Let* $e = e(z,n)$ *denote the generalized eigenvectors constructed earlier for any complex* z *with* $\operatorname{Im} z < 0$ *and any* n *in* $\mathfrak{N}$. *Denote by* $e_{\mathcal{K}}$ *the projection of* e *into* $\mathcal{K}$; *that is,* $e_{\mathcal{K}}$ *is* e *minus its* $\mathcal{D}_-$ *and* $\mathcal{D}_+$ *components. Using the projections* P_- *and* P_+, *which remove the* $\mathcal{D}_-$ *and* $\mathcal{D}_+$ *components, we can write* $e_{\mathcal{K}}$ *as*

(2.50) $$e_{\mathcal{K}} = P_-e + P_+e - e .$$

ASSERTION: $e_{\mathcal{K}}(z,n)$ *can be continued as an analytic function of* z *whose values lie in* $\mathcal{H}$ *into* $-i\rho_0(B^a)$. *In particular, if the resolvent of* B^a *is meromorphic in the whole complex plane, so is* $e_{\mathcal{K}}(z,n)$.

PROOF. It is clear from the definition of

$$g = P_+^a f ,$$

where f is given by formula (2.39), that $e_{\mathcal{K}}$ can be written in the form

$$e_{\mathcal{K}} = g + m \tag{2.51}$$

where m is that element of $\mathcal{H}$ whose incoming representer is

$$m(s) = \begin{cases} (1-\xi)e^{-izs}n & \text{for} \quad -a < s < \infty \\ 0 & \text{elsewhere} ; \end{cases} \tag{2.52}$$

here ξ is defined in (2.38). It is obvious from (2.52) that m is an entire analytic function of z whose values lie in L_2; therefore m is an entire analytic function of z whose values lie in $\mathcal{H}$.

It follows from (2.45) that g can be continued analytically into $-i\rho_o(B^a)$ as an analytic function of z whose values lie in $\mathcal{H}$; it follows by (2.51) that the same is true of $e_{\mathcal{K}}(z)$. This completes the proof of Theorem 2.10. ∎

One of the consequences of Theorem 2.10, see Section 7, is that the generalized eigenfunctions which are the concrete realizations of the generalized eigenvectors e(z, n) also can be continued analytically into $-i\rho(B^a)$. In particular if this set includes the entire real axis, the generalized eigenfunction, which we denote again by $e(\sigma, n)$, σ real, can be used to obtain in a concrete fashion the incoming spectral representation of A in $\mathcal{H}$. We now indicate how this is done in the case of spectral multiplicity one, that is when $\mathcal{N}$ corresponds to the complex numbers.

Let f be some element of $\mathcal{H}$ and as before denote by f_- its incoming translation representation. Then the incoming spectral representation of f is given by

$$\tilde{f}_-(\sigma) = \frac{1}{\sqrt{2\pi}} \int f_-(s)\, e^{i\sigma s}\, ds \,. \tag{2.53}$$

Suppose f is such that $f_-(s)$ vanishes for s large negative. Then it follows from (2.53) that

(2.54) $$\tilde{f}_-(\sigma) = \lim_{\tau \uparrow 0} \frac{1}{\sqrt{2\pi}} \int f_-(s)\, e^{i\sigma s + \tau s}\, ds \ .$$

Introducing

$$z = \sigma + i\tau$$

this can be rewritten symbolically as

(2.55) $$f_-(\sigma) = \lim_{\tau \uparrow 0} \frac{1}{\sqrt{2\pi}} (f_-, e^{-izs}) \ .$$

For elements f and g of $\mathcal{H}$, it follows from the unitarity of the incoming representation that

(2.56) $$(f, g) = (f_-, g_-) \ .$$

We shall show in Section 7 that this relation can be extended to pairs f and g when g is the concrete realization of an element of $\overline{\mathcal{H}}$. Using (2.56) with $g_-(s) = e^{-isz}$ we can rewrite (2.55) as follows:

(2.57) $$\tilde{f}_-(\sigma) = \lim_{\tau \uparrow 0} \frac{1}{\sqrt{2\pi}} (f, e(z)) \ .$$

In Section 7 we shall show, using Theorem 2.10, that

$$\lim_{\tau \uparrow 0} (f, e(z)) = (f, e(\sigma)) \ ;$$

so (2.57) can be rewritten as

(2.58) $$\tilde{f}_-(\sigma) = \frac{1}{\sqrt{2\pi}} (f, e(\sigma)) \ .$$

Although originally derived for f whose incoming representations are zero near $-\infty$, formula (2.58) holds because of unitarity for all f in $\mathcal{H}$ if suitably interpreted.

Appendix 1 to Section 2: The generalized condition (iii).

In some applications of the above theory the U-translates of $\mathcal{D}_-$ and those of $\mathcal{D}_+$ are not dense in $\mathcal{H}$ as required by condition (iii) of (2.1), but rather they are dense in a subspace $\mathcal{H}_0$ of $\mathcal{H}$, that is

$$\overline{\vee U(t)\mathcal{D}_-} = \mathcal{H}_0 = \overline{\vee U(t)\mathcal{D}_+} \ . \tag{2.59}$$

We note that this subspace is obviously invariant under U(t). Thus if $\mathcal{D}_-$ and $\mathcal{D}_+$ also satisfy conditions (i) and (ii) of (2.1), then the theory applies directly to the restriction of U to $\mathcal{H}_0$.

The following lemma sheds light on the nature of the subspace that could appear in condition (2.59).

LEMMA 2.11. *Suppose* $\mathcal{D}_-$ *satisfies conditions* (i)_ *and* (ii) *of* (2.1) *and set*

$$\mathcal{H}_- = \overline{\vee U(t)\mathcal{D}_-} \ .$$

Then U(t) *restricted to* $\mathcal{H}_-$ *has an absolutely continuous spectrum.*

PROOF. As before $\mathcal{H}_-$ is left invariant by U so that $\mathcal{D}_-$ is an incoming subspace for U restricted to $\mathcal{H}_-$ satisfying conditions (i)-(iii) of (2.1). The translation representation theorem therefore applies and shows that the spectrum of U(t) restricted to $\mathcal{H}_-$ is absolutely continuous. ∎

A similar result holds for $\mathcal{D}_+$ on

$$\mathcal{H}_+ = \overline{\vee U(t)\mathcal{D}_+} \ .$$

Denote as before the generator of U(t) by A; iA is self-adjoint over $\mathcal{H}$, and $\mathcal{H}$ can be decomposed as the orthogonal sum of three invariant subspaces:

$$\mathcal{H} = \mathcal{H}_p \oplus \mathcal{H}_c \oplus \mathcal{H}_s \ . \tag{2.60}$$

Over $\mathcal{H}_p$ the spectrum of A is pure point spectrum, over $\mathcal{H}_c$ the spectrum of A is absolutely continuous, and over $\mathcal{H}_s$ it is singular with respect to Lebesgue measure. It follows from Lemma 2.11 that the space $\mathcal{H}_o$ occurring in (2.59) must be part of $\mathcal{H}_c$.

We say that *the generalized condition* (iii) *is satisfied when* $\mathcal{H}_o$ *is as large as possible*, i.e. *equal to* $\mathcal{H}_c$.

$$\overline{VU(t)\mathfrak{D}_-} = \mathcal{H}_c = \overline{VU(t)\mathfrak{D}_+} \quad (2.61)$$

Appendix 2 to Section 2: The spaces $\mathcal{H}^\lambda$.

In an earlier part of this section we have shown that given a translation representation of $\mathcal{H}$ we can extend the space $\mathcal{H}$ to a larger space $\bar{\mathcal{H}}$ consisting of elements represented by functions which are L_2 on $\mathbf{R}_+$ and locally L_2 on $\mathbf{R}_-$. Furthermore we have shown that if we have another translation representation which is related to the first one by a scattering operator which is causal, then every element of $\bar{\mathcal{H}}$ has a representer also in this second translation representation; moreover this representer is a function which is L_2 on $\mathbf{R}_+$, locally L_2 on $\mathbf{R}_-$.

In this appendix we restrict our attention to that subspace of $\bar{\mathcal{H}}$ whose representers f in the original representation satisfy

$$\|f\|_\lambda^2 = \int_{-\infty}^{0} |f(s)|^2 e^{2\lambda s}\, ds + \int_0^\infty |f(s)|^2\, ds < \infty , \quad (2.62)$$

where λ is some given positive number. We denote this subspace by $\mathcal{H}^\lambda$, and we call the quantity (2.62) the square of the λ-*norm* of f. The advantage gained by this restriction is that we can define translation representers of elements of $\mathcal{H}^\lambda$ with respect to another translation representation related to the first representation by a scattering operator that is not necessarily causal. The disadvantage of restricting attention to $\mathcal{H}^\lambda$ is that we restrict thereby the class of eigenelements; in fact the eigenelements in $\mathcal{H}^\lambda$ correspond to the functions

$$e(s) = e^{-izs}n, \quad n \in \mathcal{N} \tag{2.63}$$

if and only if z satisfies

$$-\lambda < \operatorname{Im} z < 0 . \tag{2.64}$$

An application of this theory is presented in Section 7; it turns out that this restricted class of eigenelements is sufficiently large for our purposes.

THEOREM 2.12. *Let* C *be an operator valued* $(\mathcal{N} \to \mathcal{N})$ *function (or measure) whose support lies on* $\mathbf{R}_+$, *and suppose that*

$$c = \int_0^\infty |C(p)| e^{\lambda p} dp < \infty . \tag{2.65}$$

Then convolution with C *is a bounded operator in the* λ*-norm; in fact*

$$\|C * f\|_\lambda \le c\|f\|_\lambda , \tag{2.66}$$

where c *is the constant defined in* (2.65).

PROOF. We denote by $T(p)$ translation to the right by p:

$$(T(p)f)(s) = f(s-p) .$$

Let us denote the norm of an operator T with respect to the λ-norm for functions by $\|T\|_\lambda$. We claim that

$$\|T(p)\|_\lambda = \begin{cases} e^{\lambda p} & \text{for} \quad p > 0 \\ 1 & \text{for} \quad p < 0 . \end{cases} \tag{2.67}$$

In fact, we claim that this is an immediate consequence of the definition (2.62), and we leave the proof to the reader.

By definition, convolution is a superposition of translation:

$$C* = \int_0^\infty C(p)\,T(p)\,dp \ ,$$

so by the triangle inequality we deduce from (2.67) and (2.66) that

$$\|C*\|_\lambda \le \int_0^\infty |C(p)|\,\|T(p)\|_\lambda\,dp$$

$$\le \int_0^\infty |C(p)|\,e^{\lambda p}\,dp = c$$

this proves (2.66). ∎

THEOREM 2.13. *Let* S *be a linear operator on functions defined on* **R** *which has the following properties:*

a) S *commutes with translation*

b) S *is causal, that is* S *maps functions supported on* $\mathbf{R}_-$ *into functions supported on* $\mathbf{R}_-$.

c) S *is a contraction in the* L_2 *norm.*

CONCLUSION: S *is a contraction in the* λ*-norm defined in* (2.62).

PROOF. Given a function ℓ with finite λ-norm, define $m_\ell(p)$ by

$$m_\ell(p) = \int_p^\infty |\ell(s)|^2\,ds \ . \tag{2.68}$$

We rewrite $\|f\|_\lambda$ as

$$\|f\|_\lambda^2 = -\int_{-\infty}^{0} e^{2\lambda p}\, dm_f(p) + m_f(0) .$$

Integration by parts gives

$$\|f\|_\lambda^2 = 2\lambda \int_{-\infty}^{0} e^{2\lambda p} m_f(p)\, dp . \tag{2.69}$$

From properties (a) and (b) it follows that if $f(s) = 0$ for $s > p$, then $(Sf)(s) = 0$ for $s > p$. Now decompose f as

$$f = f_1 + f_2$$

where

$$\begin{aligned} f_1(s) &= 0 \qquad \text{for} \qquad s > p \\ f_2(s) &= 0 \qquad \text{for} \qquad s < p . \end{aligned} \tag{2.70}$$

Since S is linear

$$Sf = Sf_1 + Sf_2 .$$

Since $f_1(s) = 0$ for $s > p$, we conclude from our previous observation that

$$(Sf_1)(s) = 0 \qquad \text{for} \qquad s > p .$$

It follows from this and definition (2.68) that

$$m_{Sf}(p) = \int_p^{\infty} |Sf(s)|^2\, ds = \int_p^{\infty} |Sf_2(s)|^2\, ds$$

$$\leq \int_{-\infty}^{\infty} |Sf_2(s)|^2\, ds .$$

Using property (c) and (2.70) we get that the above is

$$\leq \int_{-\infty}^{\infty} |f_2(s)|^2\, ds = \int_p^{\infty} |f(s)|^2\, ds .$$

So we have shown that

$$m_{Sf}(p) \leq m_f(p) .$$

Substituting this into the expression (2.69) for $\|f\|_\lambda$ and the analogous expression for $\|Sf\|_\lambda$ gives the inequality

$$\|Sf\|_\lambda \leq \|f\|_\lambda$$

which proves Theorem 2.13. ∎

Appendix 3 to Section 2: Finite Blaschke products.

In this section we describe how to construct as finite products operator-valued functions which are unitary on the real axis, bounded and analytic in the lower half plane, with a prescribed finite number of zeros. Then we show that the scattering matrix associated with an orthogonal incoming-outgoing pair for which $\mathcal{D}_- \oplus \mathcal{D}_+$ has finite codimension is such a product. This material is known and is supplied here for the sake of completeness. Finally we show that the inverse of such a scattering operator satisfies the hypotheses of Theorem 2.12 in Appendix 2.

Let $\mathcal{N}$ be a Hilbert space, $n \in \mathcal{N}$, κ a complex number with $\operatorname{Im} \kappa < 0$. Denote by P the orthogonal projection onto the line spanned by n. We define the *Blaschke factor*

$$\mathcal{B}(z) = \frac{zI - \kappa P}{zI - \bar{\kappa} P} . \tag{2.71}$$

LEMMA 2.14. *$\mathcal{B}(z)$ is unitary for z real.*

PROOF. For z real we can write $\mathcal{B}(z)$ in the form

$$(A^*)^{-1} A \tag{2.72}$$

where

$$A = zI - \kappa P .$$

Furthermore A is normal; for any normal A, (2.72) is unitary, which proves Lemma 2.14. ■

Denote by $\mathcal{A}_-$ the Hardy class of vector valued functions analytic in the lower half plane:

$$\mathcal{A}_- = F L_2(R_-, \mathcal{N}) .$$

LEMMA 2.15. *Let* $a(z)$ *be in* $\mathcal{A}_-$ *and define* $b(z)$ *by*

$$b(z) = \mathcal{B}(z)\, a(z) . \tag{2.73}$$

Then

a) $b(z)$ *belongs to* $\mathcal{A}_-$

(2.74) b) $(b(\kappa), n) = 0$

c) *Every* b *in* $\mathcal{A}_-$ *which satisfies* (2.74) *can be written in the form* (2.73).

PROOF. Part (a) follows from the Paley-Wiener characterization of $\mathcal{A}_-$. To prove (b), we write

$$(b(\kappa), n) = (\mathcal{B}(\kappa)\, a(\kappa), n) = (a(\kappa), \mathcal{B}^*(\kappa)\, n) . \tag{2.75}$$

From definition (2.71),

$$\mathcal{B}^*(\kappa) = (\bar{\kappa} I - \kappa P)^{-1}(\bar{\kappa} I - \bar{\kappa} P) .$$

Since by definition of P, $Pn = n$, it follows that $\mathcal{B}^*(\kappa)\, n = 0$; (2.74) follows then from (2.75). To prove (c) we rewrite (2.73) using the definition (2.71) of $\mathcal{B}$:

(2.76) $$(zI - \bar{\kappa}P)\,b(z) = (zI - \kappa P)\,a(z)\ ,$$

a(z) to be determined. Multiplying through by P we see that

$$(z - \bar{\kappa})\,Pb(z) = (z - \kappa)\,Pa(z)\ .$$

It follows from (2.74) that $Pb(\kappa) = 0$; consequently

(2.77) $$Pa(z) = \frac{z - \bar{\kappa}}{z - \kappa}\,Pb(z)$$

is holomorphic near $z = \kappa$. Multiplying (2.76) by $(I - P)$ we obtain

$$z(I - P)\,b(z) = z(I - P)\,a(z)$$

and hence

(2.78) $$(I - P)\,a(z) = (I - P)\,b(z)\ .$$

Combining (2.77) and (2.78) we see that

(2.79) $$a(z) = Pa(z) + (I - P)\,a(z) = \mathcal{B}^{-1}b$$

is holomorphic near $z = \kappa$. On the other hand $\mathcal{B}^{-1}(z)$ is obviously bounded in the lower half-plane outside of a neighborhood of κ. It follows therefore by the Paley-Wiener theorem that a(z) belongs to $\mathfrak{A}_-$; this completes the proof of part (c). ∎

THEOREM 2.16. *Let* $\mathcal{B}$ *denote a product of factors of form* (2.71):

(2.80) $$\mathcal{B} = \Pi\mathcal{B}_j$$

where

(2.80)′ $$\mathcal{B}_j(z) = \frac{zI - \kappa_j P_j}{zI - \bar{\kappa}_j P_j}\ , \qquad \operatorname{Im}\kappa_j < 0\ ;$$

here P_j *is the projection onto the line spanned by* n_j. *Given any* a(z) *in* $\mathfrak{A}_-$, *we define* b(z) *as*

(2.81) $$b(z) = \mathcal{B}(z)\,a(z)$$

then

a) b *belongs to* $\mathcal{A}_-$

(2.82) b) $(b(\kappa_k), m_k) = 0$

where the m_k *are related to* n_k *by*

$$m_1 = n_1 , \quad m_k = \prod_1^{k-1} \mathcal{B}_j^{*-1}(\kappa_k)\, n_k \quad \textit{for } k > 1 . \tag{2.83}$$

c) *Every* b *in* $\mathcal{A}_-$ *which satisfies* (2.82) *can be written in the form* (2.81), *where the* n_k *are determined recursively from* (2.83).

A product of form (2.80) is called a *Blaschke product.*

PROOF. The theorem follows by induction on the number of factors, using Lemma 2.15. The numbers κ_k need not be distinct; but if M of the κ_k and the corresponding vectors n_k are equal, then condition (2.82) is to be interpreted as

$$\left(\left(\frac{d}{dz}\right)^j b(\kappa_k), m_k\right) = 0, \quad j = 0, 1, \cdots, M-1 ,$$

and formula (2.82) has to be modified appropriately. ■

THEOREM 2.17. *Let*

$$T_j : \mathcal{H} \to L_2(\mathbf{R}, \mathcal{N}), \quad j = 1, 2$$

be a pair of translation representations of $\mathcal{H}$; *we denote by* $\tilde{T}_j$ *the corresponding spectral representations. The scattering matrix relating* $\tilde{T}_1$ *to* $\tilde{T}_2$ *is a Blaschke product if and only if*

a) $\mathcal{D}_-^{(1)}$ *and* $\mathcal{D}_+^{(2)}$ *are orthogonal*

b) $\mathcal{K} = \mathcal{H} \ominus (\mathcal{D}_+^{(2)} \oplus \mathcal{D}_-^{(1)})$ *is finite dimensional.*

Here

$$T_1\mathcal{D}_-^{(1)} = L_2(\mathbf{R}_-), \quad T_2\mathcal{D}_+^{(2)} = L_2(\mathbf{R}_+) . \tag{2.84}$$

PROOF. We recall the semigroup $Z(t)$ defined on $\mathcal{K}$; in the outgoing representation (which in our case is T_2) $Z(t)$ acts as right shift followed by chopping off above $s = 0$. It follows from this that the eigenfunctions of Z have the form

$$\begin{cases} e^{-\mu_j s} m_j & \text{for} \quad s < 0, \qquad \operatorname{Re} \mu_j < 0 . \\ 0 & \text{for} \quad s > 0 . \end{cases} \tag{2.85}$$

If $\mathcal{K}$ is finite dimensional, the eigenfunctions of Z span $\mathcal{K}$; we assume for the sake of simplicity of presentation that there are no generalized eigenfunctions. By (2.84),

$$T_2(\mathcal{H} \ominus \mathcal{D}_+^{(2)}) = L_2(\mathbf{R}_-) ;$$

so it follows from assumption (a) that $T_2\mathcal{D}_-^{(1)}$ consists of those f in $L_2(\mathbf{R}_-)$ which are orthogonal to the functions in (2.85):

$$\int (f(s), e^{-\mu_j s} m_j)\, ds = 0 .$$

The Fourier transform $\tilde{f}$ of such an f belongs to $\mathcal{A}_-$ and satisfies

$$(\tilde{f}(i\bar{\mu}_j), m_j) = 0 . \tag{2.86}$$

Applying part (c) of Theorem 2.16 to $b = \tilde{f}$ we conclude from (2.86) that $\tilde{f}$ is of the form

$$\tilde{f} = \mathcal{B}a, \qquad a \text{ in } \mathcal{A}_- , \tag{2.87}$$

where $\mathcal{B}$ is a Blaschke product, i.e. of form (2.80), with

(2.88) $$\mathcal{B}_j(z) = \frac{zI - i\bar{\mu}_j P_j}{zI + i\mu_j P_j} .$$

Reversing this argument we see by parts (a) and (b) of Theorem 2.16 that the converse is also valid; that is, every $\tilde{\ell}$ of the form (2.87) is the image of an element of $\mathcal{D}_-^{(1)}$ under $\tilde{T}_2$.

For any f in $\mathcal{H}$ we define

(2.89) $$\tilde{\ell}^{(1)} = \mathcal{B}^{-1}\tilde{T}_2 f .$$

The mapping $f \to \tilde{\ell}^{(1)}$, being obtained from the spectral representation $\tilde{T}_2$ by multiplication by the unitary factor $\mathcal{B}^{-1}$, is also a spectral representation; it follows from the above analysis that $\mathcal{D}_-^{(1)}$ is represented by $\mathcal{A}_-$. It follows from (2.84) that the same is true of $\tilde{T}_1$; therefore these two representations differ by a constant unitary factor (see Corollary 4.1 of Chapter 2 in [14]). Hence modulo this trivial constant unitary factor, we conclude from (2.89) that $\mathcal{B}$ is the scattering matrix relating $\tilde{T}_1$ to $\tilde{T}_2$. This proves the direct part of Theorem 2.17; the proof of the converse is similar. ∎

THEOREM 2.18. *Let* T_1 *and* T_2 *be two translation representations as in Theorem* 2.17; *that is*

$$\tilde{T}_2 = \mathcal{B}\tilde{T}_1$$

where $\mathcal{B}$ *is a Blaschke product of form* (2.80) *where* $\mathcal{B}_j$ *is given by* (2.88). *Let* λ *be any number satisfying*

(2.90) $$0 < \lambda < -\operatorname{Re}\mu_j$$

for all j. *Then* $\mathcal{H}_1^\lambda$ *and* $\mathcal{H}_2^\lambda$, *the extensions of* $\mathcal{H}$ *with respect to* T_1 *and* T_2, *defined in Appendix* 2, *are isomorphic and the two* λ*-norms are equivalent.*

PROOF. Since $\mathcal{H}_1^\lambda$ and $\mathcal{H}_2^\lambda$ are defined as completions of $\mathcal{H}$ in the λ-norms, to prove the theorem it suffices to show that the λ-norms with respect to T_1 and T_2 are equivalent in $\mathcal{H}$. By property (a) the scattering operator relating T_1 to T_2 is causal; therefore it follows from Theorem 2.13 that for any f in $\mathcal{H}$ and any λ

$$\|T_2 f\|_\lambda \le \|T_1 f\|_\lambda . \tag{2.91}$$

To obtain an inequality in the opposite direction we first construct the scattering operator relating T_2 to T_1; according to the theory developed in this appendix, this operator is convolution with the Fourier inverse of the scattering matrix relating $\tilde{T}_2$ to $\tilde{T}_1$. That scattering matrix, call it $\mathcal{C}$, is the reciprocal of the scattering matrix $\mathcal{B}$ relating $\tilde{T}_1$ to $\tilde{T}_2$:

$$\mathcal{C} = \mathcal{B}^{-1} .$$

It follows from this and the definition of $\mathcal{B}$ via (2.80) and (2.88) that $\mathcal{C}^{-1}$ is a rational function, with poles at $i\bar{\mu}_j$; ∞ is a regular point, and $\mathcal{C}(\infty) = I$. Although the general case can be handled just as easily, we assume for notational convenience that all poles of $\mathcal{C}$ are simple; then $\mathcal{C}$ has the partial fraction expansion

$$\mathcal{C}(z) = I + \sum \frac{C_j}{z - i\bar{\mu}_j} . \tag{2.92}$$

Denote by $C(s)$ the Fourier inverse of $\mathcal{C}$; inverting (2.92) we get

$$C(s) = \begin{cases} 0 & \text{for} \quad s < 0 \\ \delta(s) - i \sum C_j e^{\bar{\mu}_j s} & \text{for} \quad s \ge 0 . \end{cases} \tag{2.93}$$

Since λ satisfies (2.90), it follows from (2.93) that

$$c = \int_0^\infty |C(s)| e^{\lambda p}\, dp < \infty .$$

It follows then from Theorem 2.12 that convolution with C is a bounded operator with respect to the λ-norm. Now convolution with C is the scattering operator relating T_2 to T_1; we therefore conclude that

$$\|T_1 f\|_\lambda \leq c\|T_2 f\|_\lambda \ .$$

This inequality and (2.91) together show that the λ norm with respect to T_1 and T_2 are equivalent; this proves Theorem 2.18. ∎

Being extensions of $\mathcal{H}$ with respect to equivalent norms, $\mathcal{H}_1^\lambda$ and $\mathcal{H}_2^\lambda$ can be identified and denoted as $\mathcal{H}^\lambda$. T_1 and T_2 can be extended by continuity as maps of $\mathcal{H}^\lambda$ into the space of functions with finite λ-norm.

The role of the extended space $\mathcal{H}^\lambda$ is to accommodate generalized eigenfunctions $e(z,n)$ of the unitary group $U(t)$. We define these as elements of $\mathcal{H}^\lambda$ which are represented in, say, T_2 by exponential functions:

$$(2.94)_2 \qquad T_2\, e(z,n) = e^{-izs} n \ .$$

The function on the right has finite λ-norm if

$$(2.95) \qquad -\lambda < \operatorname{Im} z < 0 \ .$$

According to (2.14), any other representation of $e(z,n)$ is again an exponential $e^{-izs}m$, $m = \mathcal{S}(z)n$. Since for the pair T_2, T_1 the scattering matrix $\mathcal{S}$ is $\mathcal{C}$, we have

$$(2.94)_1 \qquad T_1 e(z,n) = e^{-izs}\, \mathcal{C}(z) n \ .$$

We denote by $P_-^{(j)}$ and $P_+^{(j)}$ the operators which remove the $\mathbf{R}_-$ or $\mathbf{R}_+$ component respectively of $T_j \mathcal{H}^\lambda$, $j = 1, 2$.

THEOREM 2.19. *Suppose that* T_1 *and* T_2 *satisfy the hypotheses of Theorem* 2.17; *then*

$$(2.96)_- \qquad P_-^{(1)} e(z,n) - P_-^{(2)} e(z,n)$$

is a rational function of z, *with values in* $\mathcal{H}$, *whose poles are at* $i\bar{\mu}_j$; *and so is*

$$(2.96)_+ \qquad P_+^{(1)} e(z,n) - P_+^{(2)} e(z,n) \ .$$

PROOF. Since $P_-^{(1)} + P_+^{(1)} = P_-^{(2)} + P_+^{(2)} =$ identity, it follows that $(2.96)_+$ is just the negative of $(2.96)_-$; so it suffices to study $(2.96)_-$.

It follows from $(2.94)_2$ and the definition of $P_-^{(2)}$ that

$$T_2 P_-^{(2)} e = \begin{cases} 0 & \text{for} \quad s < 0 \\ e^{-izs} n & \text{for} \quad s > 0 \ . \end{cases}$$

The convolution of $T_2 P_-^{(2)} e$ with the scattering operator C yields $T_1 P_-^{(2)} e$; that is

$$T_1 P_-^{(2)} e = \int_0^\infty C(s-p)\, e^{-izp} n \, dp \ .$$

Using formula (2.93) for C we get

$$T_1 P_-^{(2)} e = \begin{cases} 0 & \text{for} \quad s < 0 \\ e^{-izs} n - i \sum C_j \int_0^s e^{\bar{\mu}_j (s-p) - izp} n \, dp & \text{for} \quad s > 0 \ . \end{cases}$$

Carrying out the integration yields for $s > 0$

$$e^{-izs} n + i \sum C_j e^{\bar{\mu}_j s} \frac{e^{-(\bar{\mu}_j + iz)s} - 1}{\bar{\mu}_j + iz} n$$

$$= e^{-izs} \left[I + \sum \frac{C_j}{-i\bar{\mu}_j + z} \right] n - \sum C_j \frac{e^{\bar{\mu}_j s}}{-i\bar{\mu}_j + z} n \ .$$

Using formula (2.92) for $\mathcal{C}(z)$ we can rewrite the above as

$$T_1 P_-^{(2)} e = \begin{cases} 0 & \text{for} \quad s < 0 \\ e^{-izs}\mathcal{C}(z)n - \sum e^{\bar{\mu}_j s} \dfrac{C_j n}{-i\bar{\mu}_j + z} & \text{for} \quad s > 0 . \end{cases} \tag{2.97}$$

On the other hand it follows from $(2.94)_1$ and the definition of $P_-^{(1)}$ that

$$T_1 P_-^{(1)} e = \begin{cases} 0 & \text{for} \quad s < 0 \\ e^{-izs}\mathcal{C}(z)n & \text{for} \quad s > 0 . \end{cases} \tag{2.98}$$

Subtracting (2.97) from (2.98) we get

$$T_1(P_-^{(1)}e(z) - P_-^{(2)}e(z)) = \begin{cases} 0 & \text{for} \quad s < 0 \\ \sum e^{\bar{\mu}_j s} \dfrac{C_j n}{-i\bar{\mu}_j + z} & \text{for} \quad s > 0 . \end{cases}$$

Clearly the right side is a rational function, with values in $\mathcal{H}$; this completes the proof of Theorem 2.19. ■

§3. A MODIFIED THEORY FOR SECOND ORDER EQUATIONS WITH AN INDEFINITE ENERGY FORM

In this section we show how the theory developed in Section 2 can be applied to motions governed by equations which are *second order in time*, that is of the form

$$u_{tt} = Lu \, , \tag{3.1}$$

where u is an element of a Hilbert space $\mathfrak{L}$ and L is a *self-adjoint operator* on $\mathfrak{L}$. Assuming that incoming and outgoing subspaces exist, the theory applies directly when L is negative; if L has a finite number of positive eigenvalues, some modifications of the theory are required.

One can easily rewrite (3.1) as a first order system by introducing the time derivative of u as a new variable:

$$u_t = v \, , \qquad v_t = Lu \tag{3.2}$$

the pair of elements $\{u, u_t\} = f(t)$ are called *data*. We define $\mathcal{H}_0$ to be the space of all data $f = \{f_1, f_2\}$ where f_1 belongs to the domain of $|L|^{\frac{1}{2}}$ and f_2 is in $\mathfrak{L}$. We can now write (3.2) more concisely in matrix notation as

$$f_t = Af \tag{3.3}$$

where

$$A = \begin{pmatrix} 0 & 1 \\ L & 0 \end{pmatrix} . \tag{3.4}$$

The data at time t is uniquely determined by its *initial data*, that is by its value at $t = 0$. The operator $U(t)$ is defined as the mapping

$$U(t) : \{u(0), u_t(0)\} \to \{u(t), u_t(t)\} .$$

The *energy* of the data f associated with equation (3.1) is defined as the quadratic function:

$$E(f) = \|u_t\|^2 - (u, Lu) , \tag{3.5}$$

where $\| \ \|$ and $(\, , \,)$ denote the norm and inner product in the Hilbert space $\mathcal{L}$. It is clear that $E(f)$ is meaningful for all f in $\mathcal{H}_0$.

THEOREM 3.1 (Conservation of energy). *The energy of a solution of* (3.1) *is independent of* t.

PROOF. If we differentiate the expression (3.5) and then make use of the equation (3.1) and the symmetry of L, we obtain

$$\begin{aligned} \frac{dE}{dt} &= (u_{tt}, u_t) + (u_t, u_{tt}) - (u_t, Lu) - (u, Lu_t) \\ &= (Lu, u_t) + (u_t, Lu) - (u_t, Lu) - (Lu, u_t) = 0 . \end{aligned} \tag{3.6}$$

Since $E_t = 0$, E is independent of t as asserted. ■

Suppose the operator L is *negative*; then we can introduce the quantity $\sqrt{E}$ as norm. We denote by $\mathcal{H}$ *the completion of* $\mathcal{H}_0$ *in the energy norm.* $\mathcal{H}$ is a Hilbert space in the energy norm and plays the role of the space denoted by $\mathcal{H}$ in Section 2.

The existence of solutions for the system (3.3) can conveniently be based on the functional calculus for the self-adjoint operator L. We denote by M the positive square root of $-L$. Then in the sense of the functional calculus we can write

$$U(t) = e^{At} = \begin{pmatrix} \cos Mt & M^{-1} \sin Mt \\ -M \sin Mt & \cos Mt \end{pmatrix} . \tag{3.7}$$

It is readily verified that $U(t)$ defines a strongly continuous group of isometries in $\mathcal{H}_0$ with respect to the energy norm. For data $f = \{f_1, f_2\}$ with f_1 in $D(L)$ and f_2 in $D(M)$ one can differentiate $U(t)f$; this gives

$$\frac{dU(t)f}{dt} = AU(t)f \ ,$$

and shows that $U(t)f$ is indeed a solution of (3.3). Extending $U(t)$ by continuity, we obtain a group of unitary operators on $\mathcal{H}$ with a skew-self-adjoint generator A.

The generator of the group U is the *closure* of A defined by (3.4) in the *core domain* consisting of data $u = \{u_1, u_2\}$ with u_1 in $D(L)$ and u_2 in $D(M)$. To verify this it is enough to show for real $\lambda \neq 0$ that the range of $(\lambda - A)$ on this domain is dense in $\mathcal{H}$. In component form we can write $(\lambda - A)u = f$ as

$$\begin{aligned} \lambda u_1 - \ \ u_2 &= f_1 \\ \lambda u_2 - Lu_1 &= f_2 \ . \end{aligned} \tag{3.8}$$

Multiplying the first equation by λ and adding this to the second we get

$$(\lambda^2 - L)u_1 = \lambda f_1 + f_2 \ .$$

Since L is negative this has a solution u_1 in $D(L)$ for any $\lambda f_1 + f_2$ in $\mathcal{L}$. If f_1 belongs to $D(M)$ then the first equation in (3.8) shows that u_2 also belongs to $D(M)$. This proves that the range of $(\lambda - A)$ on the core domain fills out $\mathcal{H}_0$, which is dense in $\mathcal{H}$.

The operators L and A are closely related; roughly speaking A is a square root of L. The following two theorems give more precise statements on this relationship.

THEOREM 3.2. *If* $-\mu^2$ *belongs to the point spectrum of* L, *then* $\pm i\mu$ *belongs to the point spectrum of* A *and vice versa.*

PROOF. Let w be an eigenvector of L with eigenvalue $-\mu^2$, $\mu \geq 0$:

$$Lw = -\mu^2 w .$$

L being negative implies that μ is positive. Using the definition (3.4) of A and the above description of D(A), it is clear that

$$f^+ = \{w, i\mu w\} \quad \text{and} \quad f^- = \{w, -i\mu w\} \tag{3.9}$$

are eigenvectors of A:

$$Af^+ = i\mu f^+ \quad \text{and} \quad Af^- = -i\mu f^- . \tag{3.10}$$

Conversely suppose that f is an eigenvector of A:

$$Af = i\mu f ,$$

or in terms of components

$$f_2 = i\mu f_1 , \quad \text{and} \quad Lf_1 = i\mu f_2 .$$

By definition of $\mathcal{H}$, f_2 belongs to $\mathcal{L}$. Hence it follows from the above relations that if $\mu \neq 0$ then both f_1 and Lf_1 belong to $\mathcal{L}$ and therefore that f_1 is in the domain of the self-adjoint operator L on $\mathcal{L}$ with

$$Lf_1 = -\mu^2 f_1 .$$

Thus if $\mu \neq 0$, $-\mu^2$ is in the point spectrum of L. The case $\mu = 0$ can not occur since $Af = 0$ and the above description of $\mathcal{D}(A)$ imply that $E(f) = 0$ so that f is a null vector in $\mathcal{H}$. ∎

A similar relation exists between the absolutely continuous spectra of L and A, which we now state without proof.

THEOREM 3.3. *Let*

$$u \leftrightarrow \phi(\mu), \qquad u \in \mathbf{R}_+ ,$$

be a unitary spectral representation of $\mathcal{L}$ *on* $L_2(\mathbf{R}_+)$ *with respect to* L, *so that for* u *in* D(L)

$$Lu \leftrightarrow -\mu^2 \phi(\mu) .$$

Then the following is a unitary spectral representation with respect to A *of* $\mathcal{H}$ *on* $L_2(\mathbf{R})$:

$$f = \{f_1, f_2\} \leftrightarrow \psi(\mu), \qquad \mu \in \mathbf{R} ,$$

where

$$\psi(\mu) = \frac{1}{\sqrt{2}} [i\mu\phi_1(|\mu|) + \phi_2(|\mu|)] ;$$

here ϕ_1 *and* ϕ_2 *are the spectral representers of* f_1 *and* f_2 *with respect to* L.[1] *Moreover under this mapping*

$$Af \leftrightarrow i\mu\psi .$$

It follows in particular from Theorem 3.3 that the continuous spectrum of A has uniform multiplicity over $\mathbf{R}$ if and only if the continuous spectrum of L has uniform multiplicity over $\mathbf{R}_+$. This remark is relevant in the choice of L in the later sections of this monograph.

We now describe the modifications that have to be made in the above construction when the operator L is not negative but has only a *finite number of eigenvectors with positive eigenvalues* (cf. [15]). We denote these eigenvectors and eigenvalues as

$$(3.11) \qquad Lq_j = \lambda_j^2 q_j , \qquad \lambda_j > 0, \ j = 1, 2, \cdots, m$$

and normalize q_j so that $\|q_j\| = 1/\sqrt{2}$. In analogy with (3.9) and (3.10) we see that the vectors

$$(3.12) \qquad f_j^{\pm} = \{q_j, \pm\lambda_j q_j\}$$

---

[1] For some f, f_1 lies in the completion of $\mathcal{L}$ with respect to $-(f, Lf)$. Such f are represented by functions ϕ for which $\sqrt{\mu}\phi$ is in L_2.

in $\mathcal{H}$ are eigenvectors of A:

$$Af_j^{\pm} = \pm\lambda_j f_j^{\pm} \tag{3.13}$$

so that

$$U(t)f_j^{\pm} = e^{\pm\lambda_j t} f_j^{\pm} . \tag{3.14}$$

The bilinear energy form, associated with the quadratic form (3.5), is

$$E(f, g) = (f_2, g_2) - (f_1, Lg_1) \tag{3.15}$$

where

$$f = \{f_1, f_2\} \quad \text{and} \quad g = \{g_1, g_2\} .$$

The following relations are easy to derive from (3.13):

$$\left.\begin{aligned} E(f_j^+, f_k^+) &= 0 \\ E(f_j^-, f_k^-) &= 0 \end{aligned}\right\} \quad \text{for all} \quad j, k ; \tag{3.16}$$

and

$$E(f_j^+, f_k^-) = \begin{cases} 0 & \text{for} \quad j \neq k \\ -\lambda_j^2 & \text{for} \quad j = k . \end{cases} \tag{3.17}$$

Denote by $\mathcal{P}$ the subspace spanned by the vectors $\{f_j^{\pm}; j = 1, \cdots, m\}$. It follows from (3.16) and (3.17) that the energy form is nondegenerate in $\mathcal{P}$ in the sense that only the zero vector of $\mathcal{P}$ is E-orthogonal to all of $\mathcal{P}$. The space $\mathcal{P}$ can be decomposed into two nonorthogonal parts

$$\mathcal{P} = \mathcal{P}_- + \mathcal{P}_+ , \tag{3.18}$$

when $\mathcal{P}_-$ is spanned by the f_j^- and $\mathcal{P}_+$ by the f_j^+.

Again we denote by $\mathcal{H}_0$ the set of all data $f = \{f_1, f_2\}$ with f_1 in the domain of $|L|^{\frac{1}{2}}$ and f_2 in $\mathcal{L}$. $\mathcal{P}$ is a subspace of $\mathcal{H}_0$ and we denote by $\mathcal{H}'_o$ the E-orthogonal complement of $\mathcal{P}$ in $\mathcal{H}_o$. Since $\mathcal{P}$ is spanned by

$$f_j^+ + f_j^- = \{2q_j, 0\} \qquad \text{and} \qquad f_j^+ - f_j^- = \{0, 2\lambda_j q_j\}$$

we see that $\mathcal{H}'_o$ consists of all data f in $\mathcal{H}_o$ such f_1 and f_2 are each $\mathcal{L}$-orthogonal to the eigenvectors q_j. It follows that E is non-negative over $\mathcal{H}'_o$. Since E is nondegenerate over $\mathcal{P}$ we also see that $\mathcal{P}$ and $\mathcal{H}'_o$ have only the zero vector in common.

We now proceed as before to construct U(t) on $\mathcal{H}'_o$ with M now equal to the square root of the *non-negative* part of $-L$. If $\mathcal{H}'_E$ denotes the E-completion of $\mathcal{H}'_o$, then as before

$$A = \begin{pmatrix} 0 & 1 \\ L & 0 \end{pmatrix}$$

restricted to $\mathcal{H}'_E$ generates a group of unitary operators $U'(t)$ in $\mathcal{H}'_E$. Under the completion, $\mathcal{H}'_E$ remains E-orthogonal to $\mathcal{P}$.[2]

We denote the direct sum of $\mathcal{P}$ and $\mathcal{H}'_E$ by $\mathcal{H}$:

$$\mathcal{H} = \mathcal{P} \oplus \mathcal{H}'_E . \tag{3.19}$$

We define the action of U(t) in $\mathcal{H}$ to be the combined action of U in $\mathcal{P}$ given by (3.14) and that of U' in $\mathcal{H}'_E$; then U is unitary with respect to the energy form.

We denote by Q' the E-orthogonal projection of $\mathcal{H}$ onto $\mathcal{H}'_E$; in terms of the biorthogonal functions $\{f_j^\pm\}$, Q' is given by

$$Q'f = f + \sum a_j f_j^+ + \sum b_j f_j^- \tag{3.20}$$

where

$$a_j = E(f, f_j^-)/\lambda_j^2 \qquad \text{and} \qquad b_j = E(f, f_j^+)/\lambda_j^2 .$$

---

[2]It may happen that L has a null vector q_o, $Lq_o = 0$; in this case $f = \{q_o, 0\}$ in $\mathcal{H}'_o$ is zero in $\mathcal{H}'_E$ and $g = \{0, q_o\}$ is annihilated by A.

Since $\mathcal{P}$ and $\mathcal{H}'_E$ are both invariant subspaces for U, it follows that

$$Q'U(t) = U(t)Q' . \tag{3.21}$$

This is not the end of the story however, since we have not yet introduced the distinguished subspaces $\mathcal{D}_-$ and $\mathcal{D}_+$ of $\mathcal{H}$. In the case of the automorphic wave equation $\mathcal{D}_-$ and $\mathcal{D}_+$ do not lie in $\mathcal{H}'_E$. The rest of this section is devoted to describing the modifications in the theory of Section 2 required by this fact.

Let $\mathcal{D}_-$ and $\mathcal{D}_+$ be a pair of subspaces of $\mathcal{H}$ which satisfy conditions (i) and (ii) of (2.1) and which are *closed* in the sense of *weak sequential convergence relative to the form* E:

$$E-\text{weak}\lim f_n = f$$

means that

$$\lim E(f_n, g) = E(f, g) \tag{3.22}$$

for all g in $\mathcal{H}$.

A set in $\mathcal{H}$ is called *weakly closed* if it contains all of its E-weak limits.

We begin by proving two basic relations between $\mathcal{D}_\pm$ and $\mathcal{P}$.

LEMMA 3.4. $\mathcal{P} \cap \mathcal{D}_- = \{0\} = \mathcal{P} \cap \mathcal{D}_+$.

PROOF. Suppose that some element p of $\mathcal{P}$,

$$p = \sum a_j f_j^+ + \sum b_j f_j^- \tag{3.23}$$

belongs to, say, $\mathcal{D}_+$. We assume that each term in (3.23) stands for an eigenfunction whose eigenvalue is different from all others in (3.23); we can achieve this by lumping together as one term all eigenfunctions which have a common eigenvalue. By property $(2.1)_i$, U(t)p also belongs to $\mathcal{D}_+$ for each $t \geq 0$. Using (3.14) we see that

$$U(t)p = \sum a_j e^{\lambda_j t} f_j^+ + \sum b_j e^{-\lambda_j t} f_j^- . \tag{3.24}$$

Take $U(t)p$ for as many different values $t_1, t_2, \cdots,$ of t as there are terms in (3.24). Since the λ_j's are all different in (3.24), it is possible to solve for each vector $a_j f_j^+$ and $b_j f_j^-$ as linear combinations of the elements $\{U(t_k)p\}$. Since all the $U(t_k)p$ belong to $\mathcal{D}_+$, so do $a_j f_j^+$, $b_j f_j^-$. Again applying (3.14) we see that these vectors $a_j f_j^+$ and $b_j f_j^-$ belong to $U(t)\mathcal{D}_+$ for all positive t and therefore to the intersection of these sets; this is contrary to property $(2.1)_{ii}$ unless these vectors are zero, in which case $p = 0$. ∎

LEMMA 3.5. *$\mathcal{D}_+$ is E-orthogonal to $\mathcal{P}_-$ and $\mathcal{D}_-$ is E-orthogonal to $\mathcal{P}_+$.*

PROOF. Let d be any vector in, say $\mathcal{D}_+$; decompose d according to (3.19) as the sum of a vector d' in $\mathcal{H}'_E$ and p in $\mathcal{P}$.

$$d = d' + p ; \tag{3.25}$$

p is of the form (3.23).

We wish to show that all of the coefficients a_j in (3.23) are zero. To prove this we argue indirectly, assuming the contrary. We then apply $U(t)$ to (3.25), using the expression (3.24) for $U(t)p$ we get

$$U(t)d = U(t)d' + \sum a_j e^{\lambda_j t} f_j^+ + \sum b_j e^{-\lambda_j t} f_j^- . \tag{3.26}$$

Denote the largest λ_j for which $a_j \neq 0$ by λ_m. Multiplying (3.25) by $e^{-\lambda_m t} a_m^{-1}$, this becomes

$$a_m^{-1} e^{-\lambda_m t} U(t)d = f_m^+ + r(t) ,$$

when the remainder r is of the form.

$$r = \sum c_j e^{-a_j t} f_j^{\pm} + \text{const. } e^{-\lambda_m t} U(t) d' \; ;$$

here all of the a_j are *positive*.

We claim that $r(t)$ tends E-weakly to zero as $t \to \infty$. This is clearly true of each term $e^{-a_j t} f_j^{\pm}$. It is also true of $e^{-\lambda_m t} U(t) d'$. In fact

$$E(U(t)d', f) = E(U(t)d', Q'f)$$

and since E is positive on $\mathcal{H}'_E$ and U is E-unitary we see that

$$|E(U(t)d', f)|^2 \leq E(d')E(Q'f) .$$

It follows that

$$f_m^+ = \underset{t\to\infty}{E \text{ weak lim}}\; a_m^{-1} e^{-\lambda_m} U(t) d \; . \tag{3.27}$$

Since d belongs to $\mathcal{D}_+$, property (i) of (2.1) implies that $U(t)d$ lies in $\mathcal{D}_+$ for all $t > 0$. Because $\mathcal{D}_+$ is E-weakly closed, (3.27) shows that f_m^+ belongs to $\mathcal{D}_+$; this is contrary to Lemma 3.4 and hence our earlier assumption about not all a_j being zero must be false. It follows from this that p, as given by (3.23) belongs to $\mathcal{P}_-$:

$$d = d' + p_- \; . \tag{3.28}$$

According to (3.16) any element of $\mathcal{P}_-$ is E-orthogonal to $\mathcal{P}_-$. Since any element of $\mathcal{H}'_E$ is also E-orthogonal to $\mathcal{P}_-$, this proves that d is E-orthogonal to $\mathcal{P}_-$. The second half of the lemma, asserting the E-orthogonality of $\mathcal{D}_-$ to $\mathcal{P}_+$, can be proved analogously. ∎

As mentioned earlier, for the automorphic wave equation $\mathcal{D}_-$ is *not* E-orthogonal to $\mathcal{P}_-$ nor $\mathcal{D}_+$ to $\mathcal{P}_+$ so that neither $\mathcal{D}_-$ nor $\mathcal{D}_+$ belong to $\mathcal{H}'_E$ and hence they are not suitable incoming and outgoing spaces in the sense of Section 2. Instead we introduce *two other pairs of subspaces of* $\mathcal{H}'_E$, derived from $\mathcal{D}_-$ and $\mathcal{D}_+$, which can serve as incoming and outgoing subspaces for U(t) restricted to $\mathcal{H}'_E$. Recalling that Q' is the E-orthogonal projection of $\mathcal{H}$ onto $\mathcal{H}'_E$, we define

$$\mathcal{D}'_- = Q'\mathcal{D}_- , \quad \mathcal{D}''_- = \mathcal{D}_- \cap \mathcal{H}'_E ;$$

$$\mathcal{D}'_+ = Q'\mathcal{D}_+ , \quad \mathcal{D}''_+ = \mathcal{D}_+ \cap \mathcal{H}'_E .$$

REMARK 3.6. It is clear from these definitions that $\mathcal{D}''_+ \subset \mathcal{D}'_+$. Further $\mathcal{D}'_+$ consists of elements $d' = Q'd$, d in $\mathcal{D}_+$; such a d' is of the form (3.28). By Lemma 3.4, d' will belong to $\mathcal{D}_+$ and hence to $\mathcal{D}''_+$ if and only if $p_- = 0$. Since p_- belongs to $\mathcal{P}_-$, it follows from the biorthogonality relations (3.17) that $p_- = 0$ if and only if p_-, and hence d, is E-orthogonal to $\mathcal{P}_+$. Now $\mathcal{P}_+$ is m-dimensional. Hence the codimension of $\mathcal{D}''_+$ in $\mathcal{D}_+$ is at most m and since $\mathcal{D}'_+ = Q'\mathcal{D}_+$ and $\mathcal{D}''_+ = Q'\mathcal{D}''_+$, the same is true of the codimension of $\mathcal{D}''_+$ in $\mathcal{D}'_+$. Similarly the codimension of $\mathcal{D}''_-$ in $\mathcal{D}'_-$ is at most m.

LEMMA 3.7. *Both pairs of subspaces* $\mathcal{D}'_-$, $\mathcal{D}'_+$ *and* $\mathcal{D}''_-$, $\mathcal{D}''_+$ *satisfy conditions* (i) *and* (ii) *of* (2.1) *and each of these subspaces is closed in* $\mathcal{H}'_E$.

PROOF. Since $\mathcal{D}_-$ and $\mathcal{D}_+$ satisfy condition (i), the analogous statement for $\mathcal{D}'_-$ and $\mathcal{D}'_+$ follows from the observation that U(t) and Q' commute; condition (i) for $\mathcal{D}''_-$ and $\mathcal{D}''_+$ follows from the fact that U(t) leaves $\mathcal{H}'_E$ invariant. Since $\mathcal{D}''_+ \subset \mathcal{D}_+$, condition (ii) for $\mathcal{D}''_+$ is a consequence of (ii) for $\mathcal{D}_+$.

Suppose that g' belongs to $\Lambda U(t)\mathcal{D}'_+$ which is contained in $\mathcal{D}'_+$ by (i). Then for each $t > 0$ there is an element h'_t in $\mathcal{D}'_+$ such that

$$U(t)\, h'_t = g' . \tag{3.29}$$

Making use of the decomposition (3.28) we can write

$$g' = g - p \qquad \text{and} \qquad h'_t = h_t - p_t$$

where g, h_t belong to $\mathcal{D}_+$ and p, p_t belong to $\mathcal{P}$. We can now rewrite the relation (3.29) as

$$U(t)\,h_t - g = U(t)\,p_t - p\ .$$

For $t > 0$ the left side lies in $\mathcal{D}_+$, the right side in $\mathcal{P}$ and hence, by Lemma 3.4, both sides equal 0. In particular

$$U(t)\,h_t = g\,, \qquad t > 0\ ,$$

and since condition (ii) holds for $\mathcal{D}_+$ we see that $g = 0$. It follows that $g' = Q'g = 0$ and hence that $\mathcal{D}'_+$ satisfies condition (ii); a similar argument shows that this is also true of $\mathcal{D}'_-$. Moreover $\mathcal{D}''_-$ and $\mathcal{D}''_+$ are each the intersection of E-weakly closed subspaces. They are therefore E-weakly closed and since they are contained in $\mathcal{H}'_E$ they are *a fortiori* strongly closed. As noted above, the quotient spaces $\mathcal{D}'_-/\mathcal{D}''_-$ and $\mathcal{D}'_+/\mathcal{D}''_+$ are finite dimensional and therefore $\mathcal{D}'_-$ and $\mathcal{D}'_+$ are also closed. ∎

LEMMA 3.8. *The mappings by* Q' *of* $\mathcal{D}_+$ *into* $\mathcal{D}'_+$ *and* $\mathcal{D}_-$ *into* $\mathcal{D}'_-$ *are one-to-one, onto and energy preserving.*

PROOF. The mappings are by definition onto. To prove that the map is one-to-one we have to show that no nonzero element of $\mathcal{D}_\pm$ is annihilated by Q'. This is so because the null space of Q' consists of $\mathcal{P}$ and we have already shown in Lemma 3.4 that $\mathcal{P} \cap \mathcal{D}_\pm = \{0\}$.

To prove the energy preserving property we use (3.28) according to which d in $\mathcal{D}_+$ is of the form: $d = d' + p_-$ where d' belongs to $\mathcal{D}'_+$ and p_- to $\mathcal{P}_-$. Thus

$$E(d) = E(d') + E(p_-, d') + E(d', p_-) + E(p_-)\ .$$

It follows from (3.16) that $E(p_-) = 0$ and since d' lies in $\mathcal{H}'_E$ we also have $E(d', p_-) = 0$. We conclude that

$$E(d) = E(d') \tag{3.30}$$

as stated. ∎

REMARK. Since E is positive on $\mathcal{D}'_{\pm} \subset \mathcal{H}'_E$, it follows from (3.30) that E is also positive on $\mathcal{D}_{\pm}$.

The operator A will in general have purely imaginary eigenvalues in addition to the $\{\pm \lambda_j\}$. The nonzero ones correspond, as in Theorem 3.2, to negative eigenvalues of L. There can also be null vectors; for instance if $Lq_0 = 0$, then $f = \{0, q_0\}$ in $\mathcal{H}'_E$ is such a null vector since $Af = \{q_0, 0\}$ is zero in $\mathcal{H}'_E$. In any case the corresponding eigenvectors are E-orthogonal to the $\{f_j^{\pm}\}$ and hence belong to $\mathcal{H}'_E$. We denote by $\mathcal{H}'_p$ the space spanned by these eigenvectors and by $\mathcal{H}'_c$ the E-orthogonal complement of $\mathcal{H}'_p$ in $\mathcal{H}'_E$.

We now make the further assumption – which will be verified in Section 6 for the automorphic wave equation – *that* $\mathcal{D}''_-$ *and* $\mathcal{D}''_+$ *satisfy condition* (2.51):

$$\overline{\mathrm{V}U(t)\mathcal{D}''_-} = \mathcal{H}'_c = \overline{\mathrm{V}U(t)\mathcal{D}''_+} \,. \tag{3.31}$$

It follows from Lemma 2.11 that

$$\overline{\mathrm{V}U(t)\mathcal{D}'_{\pm}} \subset \mathcal{H}'_c$$

and since $\mathcal{D}'_{\pm} \supset \mathcal{D}''_{\pm}$ we conclude that $\mathcal{D}'_-$ and $\mathcal{D}'_+$ also satisfy (3.31).

According to the Translation Representation Theorem 2.1, there exist two incoming translation representations for $U(t)$ over $\mathcal{H}'_c$, one with respect to $\mathcal{D}'_-$ and the other with respect to $\mathcal{D}''_-$. Similarly these are two outgoing translation representations, one with respect to $\mathcal{D}'_+$ and the other with respect to $\mathcal{D}''_+$. We denote these various representers of f in $\mathcal{H}'_c$ by

$$f'_-, \; f''_-, \; f'_+, \; f''_+ \,,$$

respectively.

Since $\mathcal{D}'_-$, $\mathcal{D}''_-$ are incoming and $\mathcal{D}'_+$, $\mathcal{D}''_+$ are outgoing subspaces, two obvious candidates for the role of scattering operator, in the sense of Section 2, are

(3.32)′ $$S' : \ell'_- \to \ell'_+$$

and

(3.32)″ $$S'' : \ell''_- \to \ell''_+ \,.$$

As we shall see, S' is the most natural choice. We also define the operators S_- and S_+ as

(3.32)$_\pm$
$$S_- : \ell''_- \to \ell'_-$$
$$S_+ :: \ell'_+ \to \ell''_+ \,.$$

Clearly the following relation holds between these operators.

(3.33) $$S'' = S_+ S' S_- \,.$$

The operators S_- and S_+ can also be regarded as scattering operators. In fact, if we apply Lemma 2.2 to the space $\mathcal{H}'_c$ and the subspaces $\mathcal{D}'_-$ and $\mathcal{D}'_+$ respectively, we see that S_- and S_+ are scattering operators associated with the pairs of subspaces $(\mathcal{D}''_-, \mathcal{D}'^{\perp}_-)$ and $(\mathcal{D}'^{\perp}_+, \mathcal{D}''_+)$ respectively. Observe that both pairs consist of subspaces *orthogonal to each other*; it follows therefore that both S_- and S_+ are *causal*, in the terminology introduced in Section 2.

We now make the final assumption that $\mathcal{D}_-$ *and* $\mathcal{D}_+$ *are* E-*orthogonal to each other*. It follows that $\mathcal{D}''_-$ and $\mathcal{D}''_+$ are also E-orthogonal to each other; this implies that the scattering operator S'' is causal. On the other hand it does not follow that $\mathcal{D}'_-$ and $\mathcal{D}'_+$ are E-orthogonal; in fact in the case of the automorphic wave equation $\mathcal{D}'_-$ and $\mathcal{D}'_+$ are definitely not E-orthogonal.

Denote by $\mathcal{S}'$, $\mathcal{S}''$, $\mathcal{S}_+$ and $\mathcal{S}_-$ the scattering matrices obtained by Fourier transformation from the corresponding scattering operators. It follows from (3.33) that

(3.34) $$\mathcal{S}'' = \mathcal{S}_+ \mathcal{S}' \mathcal{S}_- \,.$$

According to Theorem 2.5, $\mathcal{S}_-$, $\mathcal{S}''$ and $\mathcal{S}_+$ are all multiplications by operator valued functions which are bounded and analytic in the lower half plane. As the next theorem shows, $\mathcal{S}_-$ and $\mathcal{S}_+$ have an especially simple structure.

DEFINITION. We say that λ_j is $-$ [or $+$] *relevant* if $E(\mathcal{D}_-, f_j^-) \neq 0$ [or if $E(\mathcal{D}_+, f_j^+) \neq 0$]. If λ_j is both $+$ and $-$ relevant we simply say that it is *relevant*.

THEOREM 3.9. *$\mathcal{S}_-$ and $\mathcal{S}_+$ are Blaschke products. The zeros of $\mathcal{S}_-[\mathcal{S}_+]$ are precisely the points $-i\lambda_j$ for the $-$relevant [$+$ relevant] λ_j's.*

PROOF. Since scattering matrices are unitary on the real axis, their values at conjugate complex points are related by the Schwarz reflection principle. A zero at $-i\lambda_j$ in the lower half plane corresponds to a pole at $i\lambda_j$ in the upper half plane. To locate the poles of a scattering matrix we make use of Theorems 2.8 and 2.9 which identify the singularities of the analytic continuation of a scattering matrix with $-i$ times the eigenvalues of the infinitesimal generator B of the semigroup Z associated with the scattering system. According to (2.25)

$$Z(t) = P_+ U(t) P_-$$

where P_+ and P_- are orthogonal projections which remove the outgoing and incoming subspaces and where Z acts on the subspace $\mathcal{K} = P_- P_+ \mathcal{H}$. For the scattering matrix $\mathcal{S}_-$ the incoming subspace is $\mathcal{D}''_-$ and the outgoing subspace is $\mathcal{D}'^{\perp}_-$; the subspace $\mathcal{K}_-$ on which Z_- acts is therefore

$$\mathcal{K}_- = \mathcal{D}'_- \ominus \mathcal{D}''_- . \tag{3.35}$$

The action of Z_- on $\mathcal{K}_-$ is given by

(3.36) $$Z_-(t) = Q'_- U(t), \qquad t \geq 0 ,$$

where Q'_- is the E-orthogonal projection on $\mathcal{D}'_-$.

It follows from the Remark 3.6 and (3.35) that $\mathcal{K}_-$ is at most m-dimensional, where m is the number of positive eigenvalues of L. Consequently the generator B_- of the operators $Z_-(t)$ has a spectrum consisting of at most m eigenvalues. In the following theorem we pinpoint these eigenvalues; the proof of Theorem 3.10 will be found in the appendix to this section.

THEOREM 3.10. *The eigenvectors g_j of Z_- in $\mathcal{K}_-$ can be expressed in terms of the E-orthogonal projections Q' and Q_- of $\mathcal{H}$ onto $\mathcal{H}'_E$ and $\mathcal{D}_-$, respectively, and the eigenvectors f_j^- of A defined in* (3.12), *as follows:*

(3.37) $$g_j = Q' Q_- f_j^- .$$

The set of g_j's span $\mathcal{K}_-$ and

(3.38) $$Z(t) g_j = e^{-\lambda_j t} g_j .$$

Only the g_j's corresponding to – relevant λ_j's are nonzero.

It follows from Theorem 3.10 that the eigenvalues of $Z_-(t)$ over $\mathcal{K}_-$ are of the form $e^{-\lambda_j t}$ and hence that the eigenvalues of B_- over $\mathcal{K}_-$ are of the form $-\lambda_j$.

We now appeal to Theorems 2.8 and 2.17 to conclude that the scattering matrix $\mathcal{S}_-$ is a Blaschke product with poles at most at the $\{i\lambda_j\}$. If we use the more precise result quoted in the Remark following Theorem 2.9, we can assert that $\mathcal{S}_-$ has a pole at $i\lambda_j$ if and only if λ_j is – relevant. The zeros of $\mathcal{S}_-$ occur at the conjugate points.

This completes the proof of Theorem 3.9 for $\mathcal{S}_-$. The proof for $\mathcal{S}_+$ is essentially the same if we use the concept of time reversal, explained below. ∎

For a group U(t) defined with the aid of a second order equation and acting on a Hilbert space $\mathcal{H}$ of Cauchy data $\{u, v\}$, one can define the *time reversal operator* R as follows:

$$R : \{u, v\} \to \{u, -v\} . \tag{3.39}$$

R is a E-unitary involution of $\mathcal{H}$; it anticommutes with A:

$$RA = -AR \tag{3.40}$$

and hence

$$R\, U(t) = U(-t) R . \tag{3.41}$$

In addition, by (3.12) we have $Rf_j^+ = f_j^-$. Under the action of R an incoming subspace becomes outgoing and vice versa; in particular the roles of $\mathcal{K}_-$, Z_- and S_- are taken over by $\mathcal{K}_+$, Z_+^* and S_+^{-1}, respectively. Hence the assertion of Theorem 3.9 follows for $F S_+^{-1} F^{-1}$ under time reversal; here F is the Fourier transform. Since R changes s into $-s$ in the translation representation, it changes z into $-z$ in the spectral representation. Combining this with the relation

$$\mathcal{S}_+^{-1}(z) = \mathcal{S}_+^*(\bar{z}) ,$$

obtained by analytic continuation from the real axis, we see that $\mathcal{S}_+(z)$ has a pole at z_0 if and only if $F S_+^{-1} F^{-1}$ considered above has a pole at $-\bar{z}_0$. This proves the $\mathcal{S}_+$ assertion in Theorem 3.9.

It should be noted that the poles of $\mathcal{S}_-$ need not consist of the same subset of the $\{-i\lambda_j\}$ as the poles of $\mathcal{S}_+$. We now make the further assumption:

$$R\mathcal{D}_- = \mathcal{D}_+ ; \tag{3.42}$$

this assures that this is so. In fact it follows from (3.42) that $\mathcal{K}_-$ and $\mathcal{K}_+$ as well as the action of Z_- and Z_+^* are unitarily equivalent under R, and therefore, since the poles and zeros of $\mathcal{S}_-$ and $\mathcal{S}_+$ are purely imaginary, $\mathcal{S}_-$ and $\mathcal{S}_+$ will have the same set of poles and zeros.

We have already remarked that $\mathcal{S}'$ is the natural choice for the scattering operator in this theory. From (3.34) we get

$$\mathcal{S}' = \mathcal{S}_+^{-1} \mathcal{S}'' \mathcal{S}_-^{-1} . \tag{3.43}$$

In Section 7 we shall show for the automorphic wave equation that $\mathcal{S}''$ has a meromorphic extension into the whole complex plane. The relation (3.43) then shows that $\mathcal{S}'$ is meromorphic in the whole plane, having at most a finite number of poles in the lower half plane. It would seem from Theorem 3.10 that the poles due to $\mathcal{S}_-^{-1}$ and $\mathcal{S}_+^{-1}$ reinforce each other to create poles of the second order in the lower half plane. As we now show, $\mathcal{S}''$ has zeros at these points which cancel the poles of either one, but not both, of the operators $\mathcal{S}_-^{-1}$ and $\mathcal{S}_+^{-1}$.

THEOREM 3.11. *$S''S_-^{-1}$ and $S_+^{-1}S''$ are both causal scattering operators and hence $\mathcal{S}''\mathcal{S}_-^{-1}$ and $\mathcal{S}_+^{-1}\mathcal{S}''$ are holomorphic in the lower half plane.*

PROOF. The operator $S''S_-^{-1}$ maps the $\mathcal{D}'_-$ translation representation into the $\mathcal{D}''_+$ translation representation. In this case the roles of incoming and outgoing subspaces are played by $\mathcal{D}'_-$ and $\mathcal{D}''_+$, respectively. Causality requires that $\mathcal{D}'_-$ and $\mathcal{D}''_+$ be E-orthogonal. Since $\mathcal{D}'_-$ is the projection of $\mathcal{D}_-$ into $\mathcal{H}'_E$, each element d' of $\mathcal{D}'_-$ is of the form

$$d' = d - p ,$$

where d belongs to $\mathcal{D}_-$ and p to $\mathcal{P}$. Since $\mathcal{D}''_+ = \mathcal{D}_+ \cap \mathcal{H}'_E$, it is clear that $\mathcal{D}''_+$ is E-orthogonal to both $\mathcal{D}_-$ and $\mathcal{P}$, and hence to $\mathcal{D}'_-$. This proves causality for $S''S_-^{-1}$ and the argument for $S_+^{-1}S''$ is similar. Holomorphicity follows by Theorem 2.6. ■

In view of the fact that $\mathcal{S}_-^{-1}$ has poles at the relevant $-i\lambda_j$'s by Theorem 3.9, the above result implies

COROLLARY 3.12. *For the relevant* λ_j*'s, the scattering matrix* $\mathcal{S}''$ *has zeros at the* $-i\lambda_j$ *and when the resolvent of* B'' *is meromorphic then* B'' *has eigenvalues at the* $-\lambda_j$.

PROOF. If $\mathcal{S}'$ has a zero at $-i\lambda_j$ then it has a pole at $i\lambda_j$ and hence the assertion about B'' is a consequence of the remark following Theorem 2.9. ∎

THEOREM 3.13. $\mathcal{S}''$ *is regular in the lower half plane except for poles at the points* $\{-i\lambda_j\}$ *for the relevant* λ_j*'s.*

A proof of Theorem 3.13 will be found in the following appendix.

Appendix 1 to Section 3. Proofs of Theorems 3.10 and 3.13.

We begin by constructing the E-orthogonal projection Q_- of $\mathcal{H}$ onto $\mathcal{D}_-$, which entered into the statement of Theorem 3.10. To construct Q_- we have to decompose any f in $\mathcal{H}$ as

(3.44) $f = d_- + f'$, d_- in $\mathcal{D}_-$ and f' E-orthogonal to $\mathcal{D}_-$.

We shall do this in two steps, the first being to project f into $\mathcal{D}''_-$:

(3.45) $f = d''_- + f''$, d''_- in $\mathcal{D}''_-$ and f'' E-orthogonal to $\mathcal{D}''_-$.

In order to obtain d''_-, let Q''_- denote the E-orthogonal projection of $\mathcal{H}'_E$ onto $\mathcal{D}''_-$; the existence of Q''_- is assured by the fact that $\mathcal{D}''_-$ is a closed subspace of $\mathcal{H}'_E$ by Lemma 3.7. Then for all d in $\mathcal{D}''_-$ we have

$$E(f, d) = E(Q'f, d) = E(Q''_- Q'f, d) .$$

It is apparent from this that $d''_- = Q''_- Q'f$ furnishes us with the decomposition (3.45).

To obtain (3.44) from (3.45) we write

$$\mathcal{D}_- = \mathcal{D}''_- \oplus \mathcal{C} .$$

Using the decomposition (3.45) we see that $\mathcal{C}$ can be chosen to be E-orthogonal to $\mathcal{D}''_-$; moreover by Remark 3.6, $\mathcal{C}$ is finite dimensional. Now we modify the decomposition (3.45) as follows.

$$f = d''_- - c + f'' + c .$$

We want to choose c in $\mathcal{C}$ so that $f'' + c$ is E-orthogonal to $\mathcal{D}_-$. If we can do this we obtain the desired decomposition (3.44), with $d_- = d''_- - c$, $f' = f'' + c$. Since f'' and c are already E-orthogonal to $\mathcal{D}''_-$, the orthogonality condition reduces to

$$E(f'' + c, \mathcal{C}) = 0 ;$$

this is a determined system of inhomogeneous linear equation; a unique solution exists if the corresponding homogeneous system of equations

$$E(c, \mathcal{C}) = 0$$

has only the trivial solution $c = 0$. But this is so since $\mathcal{C}$ is a subspace of $\mathcal{D}_-$, and, as we remarked after Lemma 3.8, the form E is positive on $\mathcal{D}_-$.

PROOF OF THEOREM 3.10. By definition $\mathcal{D}'_-$ consists of elements

$$d' = Q'd, \qquad d \text{ in } \mathcal{D}_- .$$

We have already explained in Remark 3.6 that d' belongs to $\mathcal{D}''_-$ if and only if d is E-orthogonal to $\mathcal{P}_-$, that is if and only if

$$E(d, f_j^-) = 0 \qquad \text{for all } j . \tag{3.46}$$

Since d belongs to $\mathcal{D}_-$, we can replace f_j^- in the above scalar product by its component in $\mathcal{D}_-$:

$$E(d, f_j^-) = E(d, Q_- f_j^-) .$$

We now make use of Lemma 3.8, according to which Q' is an energy preserving map on $\mathfrak{D}_-$ to $\mathfrak{D}'_-$:

$$E(d, Q_- f_j^-) = E(Q'd, Q'Q_- f_j^-) = E(d', g_j)$$

where g_j is defined by (3.37): $g_j = Q'Q_- f_j^-$. It follows that condition (3.46) for $d' = Q'd$ to belong to $\mathfrak{D}''_-$ is equivalent with

(3.47) $$E(d', g_j) = 0 \qquad \text{for all } j .$$

It is clear from (3.47) that the g_j span the E-orthogonal complement of $\mathfrak{D}''_-$ in $\mathfrak{D}'_-$, which by definition is $\mathcal{K}_-$. This proves the first part of Theorem 3.10.

To prove the second part, we shall show for all d' in $\mathfrak{D}'_-$ that

(3.48) $$E(Z_-(t)\, g, d') = e^{-\lambda t} E(g, d') ,$$

where $g = g_j$ and $\lambda = \lambda_j$. Since the energy form is positive over $\mathfrak{D}'_-$, the asserted conclusion (3.38) follows by (3.36) from (3.48). In the proof of (3.48) we shall use the following facts:

a) $g = Q'Q_- f^-$, by (3.37);

b) $Z_-(t) = Q'_- U(t)$ in $\mathcal{K}_-$, by (3.36);

c) $d' = Q'd$ for $d \in \mathfrak{D}_-$, by definition of $\mathfrak{D}'_-$;

d) $E(U(t) f, g) = E(f, U(-t)\, g)$, by conservation of energy;

e) $E(Q'_- f, d') = E(f, d')$ for d' in $\mathfrak{D}'_-$, by definition of the projection Q'_-;

f) $E(Q_- f, d) = E(f, d)$ for d in $\mathfrak{D}_-$, by (3.44);

g) $E(Q'd_1, Q'd_2) = E(d_1, d_2)$ for d_1, d_2 in $\mathfrak{D}_-$, by Lemma 3.8;

h) $Q'U(t) = U(t)\, Q'$, by (3.21);

i) $U(t)\, f^- = e^{-\lambda t} f^-$, by (3.14).

Making use of the above facts we obtain the following string of identities for $t \geq 0$:

$$\begin{aligned} E(Z_{-}(t)g, d') &= E(Q'_{-}U(t)g, d') \\ &= E(U(t)g, d') = E(U(t)Q'Q_{-}f^{-}, Q'd) \\ &= E(Q'Q_{-}f^{-}, Q'U(-t)d) = E(Q_{-}f^{-}, U(-t)d) \\ &= E(f^{-}, U(-t)d) = E(U(t)f^{-}, d) \\ &= e^{-\lambda t}E(f^{-}, d) = e^{-\lambda t}E(Q_{-}f^{-}, d) \\ &= e^{-\lambda t}E(Q'Q_{-}f^{-}, Q'd) = e^{-\lambda t}E(g, d') . \end{aligned}$$

This completes the proof of (3.48) and thereby of the second part of Theorem 3.10.

It is clear from the tail end of the above string of identities that

$$E(g_j, d') = E(f_j^{-}, d) \tag{3.49}$$

and it follows from this that g_j is not zero if and only if

$$E(f_j^{-}, \mathcal{D}_{-}) \neq 0 ,$$

that is if and only if λ_j is $-$ relevant. This completes the entire proof of Theorem 3.10. ■

We conclude this appendix with

PROOF OF THEOREM 3.13. We are required to determine the singularities of $\mathcal{S}'$ in the lower half plane. In terms of the $\mathcal{D}'$ incoming and outgoing spectral representers of an element f in $\mathcal{H}'_c$, $\mathcal{S}'$ acts as multiplication.

$$\mathcal{S}'\tilde{f}'_{-} = \tilde{f}'_{+} . \tag{3.50}$$

If we choose f in $\mathcal{D}'_{-}$ then $\tilde{f}'_{-}$ will be holomorphic in the lower half plane and hence the singularities of $\tilde{f}'_{+} = \mathcal{S}'\tilde{f}'_{-}$ will be due entirely to $\mathcal{S}'$.

These singularities of $\tilde{f}'_+$ arise from the component of f'_+ with support in $\mathbf{R}_+$, that is from the E-orthogonal projection $Q'_+ f$ of f into $\mathcal{D}'_+$.

For notational reasons we relabel f above; that is we write f in $\mathcal{D}'_-$ as d'_-. Then to obtain $Q'_+ d'_-$ we compute the inner product $E(d'_-, d'_+)$ for arbitrary d'_+ in $\mathcal{D}'_+$. Now d'_- and d'_+ can be written as

$$(3.51) \qquad d'_- = d_- + p \quad \text{and} \quad d'_+ = d_+ + q$$

where by (3.28) d_- is in $\mathcal{D}_-$, d_+ in $\mathcal{D}_+$, p in $\mathcal{P}_+$ and q in $\mathcal{P}_-$. Making use of the biorthogonality relations (3.16) and (3.17), we can express p and q as

$$(3.52) \quad p = \sum \lambda_j^{-2} E(d_-, f_j^-) f_j^+ \quad \text{and} \quad q = \sum \lambda_j^{-2} E(d_+, f_j^+) f_j^- .$$

According to (3.49) and its plus counterpart

$$(3.53) \quad E(d_-, f_j^-) = E(d'_-, g_j^-) \quad \text{and} \quad E(d_+, f_j^+) = E(d'_+, g_j^+)$$

where by (3.37)

$$(3.54) \qquad g_j^- = Q'Q_- f_j^- \quad \text{and} \quad g_j^+ = Q'Q_+ f_j^+ .$$

It follows from the E-orthogonality of d_- and d_+, d'_- and q, and d'_+ and p that

$$0 = E(d_-, d_+) = E(d'_- - p, d'_+ - q) = E(d'_-, d'_+) + E(p, q) .$$

Combining this with (3.17), (3.52) and (3.53) we obtain

$$E(d'_-, d'_+) = \sum \lambda_j^{-2} E(d'_-, g_j^-) E(g_j^+, d'_+) ;$$

we conclude therefore that

$$(3.55) \qquad Q'_+ d'_- = \sum \lambda_j^{-2} E(d'_-, g_j^-) g_j^+ .$$

According to (3.36) and (3.38)

$$Z_-(t)\, g_j^- = Q'_- U(t)\, g_j^- = e^{-\lambda_j t} g_j^- \quad \text{for } t \geq 0 .$$

In terms of the incoming primed translation representation this means that

$$(\mathfrak{g}_j^-)'_-(s-t) = e^{-\lambda_j t} (\mathfrak{g}_j^-)'_-(s) \quad \text{for } s < 0$$

and hence that

$$(\mathfrak{g}_j^-)'_-(s) = e^{\lambda_j s} n_j^- \quad \text{for } s < 0 \tag{3.56}$$

for some n_j^- in $\mathfrak{N}$. Since g_j^- belongs to $\mathfrak{D}'_-$ we also have $(\mathfrak{g}_j^-)'_-(s) = 0$ for $s > 0$. Similarly in the outgoing primed translation representation

$$(\mathfrak{g}_j^+)'_+(s) = e^{-\lambda_j s} n_j^+ \quad \text{for } s > 0 \text{ and } = 0 \text{ for } s < 0 . \tag{3.57}$$

As we have observed, the singular behavior of $\mathcal{S}'$ in the lower half plane is the same as the $\mathfrak{D}'_+$ spectral representation of $Q'_+ d'_-$. Inserting (3.57) into (3.55) and taking the Fourier transform, this is seen to be

$$\sum \lambda_j^{-2} E(d'_-, g_j^-) \frac{n_j^+}{i\sigma - \lambda_j} . \tag{3.58}$$

Only those λ_j appear nontrivially in this expression for which

$$E(f_j^-, \mathfrak{D}_-) = E(g_j^-, \mathfrak{D}'_-) \neq 0$$

and $n_j^+ \neq 0$. Again $n_j^+ \neq 0$ if and only if

$$E(f_j^+, \mathfrak{D}_+) = E(g_j^+, \mathfrak{D}'_+) \neq 0 .$$

In other words only the relevant λ_j's appear nontrivially in (3.58). This completes the proof of Theorem 3.13. ■

Appendix 2 to Section 3. The semigroup Z.

The semigroup of operators Z(t) defined as in (2.25) by

$$Z(t) = P_+ U(t) P_- , \qquad (3.59)$$

plays an interesting but not a central role in the modified theory. Here as before the projections P_-, P_+ are the E-orthogonal projections on the E-orthogonal complements of $\mathcal{D}_-$ and $\mathcal{D}_+$, respectively; they can be constructed from $Q_\pm$ as $P_\pm = I - Q_\pm$. The analogue of Theorem 2.7 now reads:

THEOREM 3.14. *Let* $\mathcal{D}_-$ *and* $\mathcal{D}_+$ *be a pair of* E-*orthogonal weakly closed subspaces of* $\mathcal{H}$ *which satisfy conditions* (i) *and* (ii) *of* (2.1). *Then the* Z(t), $t \geq 0$, *form a strongly continuous semigroup of operators on*

$$\mathcal{K} = \mathcal{H} \ominus (\mathcal{D}_- \oplus \mathcal{D}_+) . \qquad (3.60)$$

Z(t) *annihilates both* $\mathcal{D}_-$ *and* $\mathcal{D}_+$.

PROOF. The proof of this theorem is the same as the proof of parts (a) and (c) of Theorem 2.7 if we utilize the fact that U is unitary with respect to the indefinite form E. ■

We denote the infinitesimal generator of Z by B and we assume that B has a pure point spectrum with finite dimensional eigenspaces. It is easy to verify for a proper eigenvector f of A with eigenvalue $\nu : Af = \nu f$, that if f is E-orthogonal to $\mathcal{D}_-$, then $P_+ f$ is an eigenvector of B with eigenvalue ν. In particular the f_j^+, the nonrelevant f_j^- and the null vectors of A correspond in this way to eigenvectors of B. The purpose of this appendix is to study another important class of eigenvectors of B, called μ-eigenvectors, and to relate these to the eigenvectors of B''.

DEFINITION. A vector f in $\mathcal{K}$ is called a μ-eigenvector of B if f is an eigenvector of B of finite index:

$$(B-\mu)^k f = 0 ,$$

with $\operatorname{Re}\mu < 0$, if f is E-orthogonal to $\mathcal{H}'_p$ and if f does not belong to $\mathcal{P}_-$.

THEOREM 3.15. *Assuming that $\mathcal{D}_-$ and $\mathcal{D}_+$ are E-orthogonal to $\mathcal{H}'_p$, there is a one-to-one correspondence between the μ-eigenvectors of* B *and the eigenvectors of* B″ *modulo the eigenvectors of* B_+.

The proof of this theorem will be broken up into several lemmas involving two other semigroups on subspaces of $\mathcal{H}'_c$. These semigroups of operators are associated in the usual way with the E-orthogonal pairs of incoming and outgoing subspaces $\mathcal{D}''_-$, $\mathcal{D}'_+$ and $\mathcal{D}'_-$, $\mathcal{D}''_+$, respectively:

$$Z^0_-(t) = P'_+ U(t) P''_- \quad \text{and} \quad Z^0_+(t) = P''_+ U(t) P'_- . \tag{3.61}$$

The theory developed in Section 2 applies to these semigroups. We denote their respective infinitesimal generators by B^0_- and B^0_+ and we shall assume the resolvents of B^0_- and B^0_+ are meromorphic in the whole complex plane. Finally we denote the associated scattering matrices by $\mathcal{S}^0_-$ and $\mathcal{S}^0_+$. As we have already shown in Theorem 3.11 both scattering matrices are causal and

$$\mathcal{S}^0_- = \mathcal{S}^{-1}_+ \mathcal{S}'' \quad \text{and} \quad \mathcal{S}^0_+ = \mathcal{S}'' \mathcal{S}^{-1}_- . \tag{3.62}$$

LEMMA 3.16. *If* f *is a* μ*-eigenvector of* B *of index* k, *then*

$$g = Q' f \tag{3.63}$$

is a nontrivial eigenvector of B^0_- *of the same index; that is*

$$(B^0_- - \mu)^k g = 0 .$$

PROOF. To simplify the exposition we consider only the case $k = 1$. We first show that f is E-orthogonal to $\mathcal{P}_-$. In fact since, by Lemma 3.5, $\mathcal{P}_-$ and $\mathcal{D}_+$ are E-orthogonal, it follows that $\mathcal{P}_-$ is in the range of P_+. Hence for f a μ-eigenvector and f_j^- in $\mathcal{P}_-$, we have

$$e^{\mu t} E(f, f_j^-) = E(Z(t) f, f_j^-) = E(P_+ U(t) f, f_j^-) = E(U(t) f, f_j^-)$$

$$= E(f, U(-t) f_j^-) = e^{\lambda_j t} E(f, f_j^-)$$

and since $\operatorname{Re} \mu < 0$, we see that $E(f, f_j^-) = 0$, as desired.

Next we note that $g \neq 0$. For if g were 0, f would belong to $\mathcal{P}$ and $\operatorname{Re} \mu < 0$ then requires that f lie in $\mathcal{P}_-$. This is contrary to our assumption on f.

We now show that g is E-orthogonal to $\mathcal{D}''_-$ and $\mathcal{D}'_+$. In fact since f is E-orthogonal to $\mathcal{P}_-$, we see from relation (3.20) that

$$f = g + q_- ,$$

where q_- lies in $\mathcal{P}_-$. By assumption f is E-orthogonal to $\mathcal{D}_-$ and hence it is automatically E-orthogonal to $\mathcal{D}''_-$ which is contained in $\mathcal{D}_-$. Moreover q_-, being an element of $\mathcal{P}$, is also E-orthogonal to $\mathcal{D}''_-$. We conclude that $g = f - q_-$ is E-orthogonal to $\mathcal{D}''_-$.

To prove that f is E-orthogonal to $\mathcal{D}'_+$, we remark that any element d'_+ in $\mathcal{D}'_+$ can be written in the form (3.28):

$$d_+ = d'_+ + p_- , \tag{3.64}$$

where d_+ belongs to $\mathcal{D}_+$ and p_- to $\mathcal{P}_-$. Thus

$$E(f, d_+) = E(g, d'_+) + E(g, p_-) + E(q_-, d'_+) + E(q_-, p_-) ;$$

making use of (3.16) and the fact that $\mathcal{H}'$ is E-orthogonal to $\mathcal{P}$, we see that

$$0 = E(f, d_+) = E(g, d'_+) .$$

This proves that g is E-orthogonal to $\mathcal{D}'_+$.

By assumption f is E-orthogonal to $\mathcal{H}_p'$ and so is q_-, since it lies in $\mathcal{P}$. Therefore g is also E-orthogonal to $\mathcal{H}_p'$. It now follows that g belongs to

$$\mathcal{K}_-^0 = \mathcal{H}_c' \ominus (\mathcal{D}_-'' \oplus \mathcal{D}_+') .$$

Finally we prove that g is an eigenvector of B_-^0:

$$B_-^0 g = \mu g . \tag{3.65}$$

Using the fact that Q' commutes with U we can write

$$Z_-^0(t)g = P_+' U(t)g = P_+' U(t)Q'f = P_+' Q'U(t)f .$$

By definition $Z(t)f = P_+U(t)f$ so that

$$U(t)f = e^t f + d_+$$

for some d_+ in $\mathcal{D}_+$. Hence

$$Z_-^0(t)g = P_+' Q'(e^{\mu t} f + d_+) = e^{\mu t} P_+' Q'f + P_+' Q'd_+ = e^{\mu t} g ,$$

from which (3.65) follows. This concludes the proof of Lemma 3.16. ∎

LEMMA 3.17. *If* g *is an eigenvector of* B_+^0 *of index* k, *then*

$$f = P_+ g \tag{3.66}$$

is a nontrivial μ*-eigenvector of* B *of the same index, provided* $\mathcal{D}_+$ *is E-orthogonal to* $\mathcal{H}_p'$.

PROOF. Again we treat only the case $k = 1$. It is easy to see that f is different from zero. Otherwise g would have to belong to $\mathcal{D}_+$; and since it also lies in $\mathcal{H}'$ it would belong to $\mathcal{D}_+''$. However this is impossible because g is in $\mathcal{K}_+^0$ which is E-orthogonal to $\mathcal{D}_+''$.

It is obvious that $f = P_+ g$ is E-orthogonal to $\mathcal{D}_+$. It is also E-orthogonal to $\mathcal{D}_-$. In fact for d_- in $\mathcal{D}_-$, we can, by (3.28), write

(3.67) $$d_- = d'_- + p_+$$

where $d'_- = Q' d_-$ lies in $\mathcal{D}'_-$ and p_+ lies in $\mathcal{P}_+$. Since

(3.68) $$f = g + d_+$$

for some d_+ in $\mathcal{D}_+$ and since $\mathcal{D}_-$ is E-orthogonal to $\mathcal{D}_+$ and $\mathcal{H}'$ to $\mathcal{P}$, we have

$$E(f, d_-) = E(g, d_-) = E(g, d'_-) = 0 ;$$

the last equality is a consequence of $\mathcal{K}^0_+$ being E-orthogonal to $\mathcal{D}'_-$. It follows that f lies in $\mathcal{K}$, the domain of Z. Moreover

$$Z(t) f = P_+ U(t)(g + d_+) = P_+ U(t) g .$$

Recalling that $Z^0_+(t) g = P''_+ U(t) g$, we see that

$$U(t) g = e^{\mu t} g + d''_+$$

for some d''_+ in $\mathcal{D}''_+$. Consequently

$$Z(t) f = P_+ U(t) g = P_+(e^{\mu t} g + d''_+) = e^{\mu t} P_+ g = e^{\mu t} f ;$$

that is, f is an eigenvector of B.

We still have to show that f is a μ-eigenvector of B. Now it follows from Theorem 2.7 that the point spectrum of B^0_+ lies in the half plane $\operatorname{Re} \mu < 0$. Since g lies in $\mathcal{H}'_c$, which is E-orthogonal to $\mathcal{H}'_p$ and since, by assumption, $\mathcal{D}_+$ is also E-orthogonal to $\mathcal{H}'_p$, we see by (3.68) that f is E-orthogonal to $\mathcal{H}'_p$. Finally if f were in $\mathcal{P}_-$, then writing d_+ as in (3.64) we can express f as

$$f = g + d_+ = g + d'_+ + p_- ,$$

and conclude that $f - p_-$ lies in $\mathcal{P}_-$ and $g + d'_+$ in $\mathcal{H}'$. Since these sets have only the zero vector in common, this means that

$$g + d'_+ = 0 .$$

Thus g belongs to the intersection of $\mathcal{K}^0_+$ and $\mathcal{D}'_+$, that is to $\mathcal{K}_+ \ominus \mathcal{D}'_-$, and is consequently both an eigenvector of B_+ and E-orthogonal to $\mathcal{D}_-$. We show in the next lemma that this is impossible. Hence, modulo this result, we have proved Lemma 3.17. ∎

LEMMA 3.18. *The eigenvectors of* B_+ *are simple and correspond through their eigenvalues* $-\lambda_j$ *to the relevant* λ_j*'s. Denote these eigenvectors by* $h_1, h_2, \cdots, h_{m'}$:

$$B_+ h_j = -\lambda_j h_j \; ; \tag{3.69}$$

and let Q'_- *denote the* E-*orthogonal projection onto* $\mathcal{D}'_-$. *Then*

$$g_j = Q'_- h_j \, , \tag{3.70}$$

is nonzero and, to within a constant factor, is equal to the g_j *eigenvector of* B_- *introduced in Theorem* 3.10; *that is*

$$B_- g_j = -\lambda_j g_j \, .$$

PROOF. If in the proof of Theorem 3.10 we reverse time, we then obtain a theorem about Z^*_+ asserting that B^*_+ has simple eigenvectors for the relevant λ_j's; the same assertion therefore holds for its adjoint B_+.

We now show that the B_+ eigenvectors h_j are not E-orthogonal to $\mathcal{D}'_-$. Since h_j lies in $\mathcal{K}_+$ and hence in $\mathcal{D}'_+$, we can, by (3.28), write

$$h_j = d_+ + q_- \, ,$$

where d_+ belongs to $\mathcal{D}_+$ and q_- to $\mathcal{P}_-$. Since h_j does not lie in $\mathcal{D}''_+$, we conclude that $q_- \neq 0$. Note that $h_j = Q' d_+$, from which it follows by

(3.20) that only the relevant f_j^-'s appear in q_-. Since $\mathfrak{D}_+$ and $\mathcal{P}_-$ are E-orthogonal we see that

$$q_- = P_+ h_j \ .$$

Again

$$U(t)\,h_j = e^{-\lambda_j t} h_j + d''_+ \tag{3.71}$$

for some d''_+ in $\mathfrak{D}''_+$ and since $P_+ = P_+P''_+$, we get

$$e^{-\lambda_j t} q_- = e^{-\lambda_j t} P_+ h_j = P_+ Z_+(t) h_j = P_+ U(t) h_j = P_+ U(t) P_+ h_j = P_+ U(t) q_- = U(t) q_- \,.$$

It follows from this that $q_- = f_j^-$. Moreover each d_- in $\mathfrak{D}_-$ can be written as in (3.67):

$$d_- = d'_- + p_+$$

for some d'_- in $\mathfrak{D}'_-$ and p_+ in $\mathcal{P}_+$. Using the fact that $d_- = P_+ d_-$, we obtain

$$E(h_j, d'_-) = E(h_j, d_-) = E(h_j, P_+ d_-) = E(P_+ h_j, d_-) = E(f_j^-, d_-) \ .$$

Since f_j^- is relevant, the right member is not identically zero and therefore h_j is not E-orthogonal to $\mathfrak{D}'_-$.

Obviously $g_j = Q'_- h_j$ is contained in $\mathfrak{D}'_-$. It is also E-orthogonal to $\mathfrak{D}''_-$ since h_j is in $\mathfrak{D}'_+$ which is E-orthogonal to $\mathfrak{D}''_-$. Thus g_j belongs to $\mathcal{K}_-$. Moreover using (3.71), we can write the following string of equalities:

$$Z_-(t)\, g_j = Q'_- U(t)\, g_j = Q'_- U(t)\, Q'_- h_j = Q'_- U(t)\, h_j = Q'_-(e^{-\lambda_j t} h_j + d''_+)$$

$$= e^{-\lambda_j t} Q'_- h_j = e^{-\lambda_j t} g_j \ .$$

This shows that g_j is indeed the B_- eigenvector $Q'Q_- f_j^-$ introduced in Theorem 3.10, at least to within a constant factor.

REMARK. Since $\mathcal{K}_-$ and $\mathcal{K}_+$ have the same dimension, it follows from Lemma 3.18 that Q'_- maps $\mathcal{K}_+$ onto $\mathcal{K}_-$ in a one-to-one fashion.

LEMMA 3.19. *The generators B^0_- and B^0_+ have the same eigenspace structure.*

PROOF. Under time reversal, Z^0_- goes into Z^{0*}_+. Since we have assumed that our system is invariant under time reversal, it follows that B^0_- and B^{0*}_+ have the same eigenspace structure. We have also assumed that the resolvent of B^0_+ is meromorphic in the entire complex plane and it follows from this that the Fredholm alternative holds. Hence B^0_+ and B^{0*}_+ have the same eigenspace structure. Combining these two conclusions we obtain the assertion of the lemma. ∎

LEMMA 3.20. *The eigenvectors of B'' are of two kinds: 1) Those which are eigenvectors of B_+; these are simple and in one-to-one correspondence with the relevant λ_j's. 2) Those which are eigenvectors of B^0_+; this correspondence is also one-to-one and onto.*

PROOF. It is clear that the B_+ eigenvectors h_j are also eigenvectors for B''. Hence part (1) is an immediate consequence of Lemma 3.18. Next suppose that g is an eigenvector for B'' which is not in the span of h_j:

$$B''g = e^{\mu t} g \ ;$$

again we consider only the case of eigenvectors of index 1. If g is not E-orthogonal to $\mathcal{D}'_-$, then Q'_-g belongs to $\mathcal{K}_-$. Using the fact that

$$U(t)g = e^{\mu t} g + d''_+$$

for some d''_+ in $\mathcal{D}''_+$, it is easy to verify that

$$Z_-(t)Q'_-g = Q'_-U(t)Q'_-g = Q'_-U(t)g = Q'_-(e^{\mu t}g + d''_+) = e^{\mu t}Q'_-g \ .$$

Hence $Q'_{-}g$ is an eigenvector for B_{-}. In view of Theorem 3.10 and Lemma 3.18, $Q'_{-}g = cg_j$ and $\mu = -\lambda_j$ for some j. Making use of the relation (3.70) we can subtract off a constant times h_j from g and obtain an eigenvector of B'' which is E-orthogonal to $\mathfrak{D}'_{-}$. Thus we may suppose to begin with that if g is not in the span of the h_j, then it is E-orthogonal to D'_{-} and hence in $\mathcal{K}^0_{+}$. In this case g is an eigenvector of B^0_{+}. Conversely it is clear that any eigenvector of B^0_{+} is also an eigenvector of B''. This proves the second part of the lemma.

The proof of Theorem 3.15 is now almost immediate. Since we have assumed the eigenspace of B to be finite dimensional, it follows from Lemma 3.17 that the eigenspaces associated with a particular eigenvalue will be finite dimensional for all of the generators. According to Lemma 3.17, P_{+} maps the eigenspace structure for B^0_{+} into the μ-eigenspace structure for B. According to Lemma 3.16, Q' maps the μ-eigenspace structure for B into the eigenspace structure for B^0_{-}. Since the eigenspace structures for B^0_{-} and B^0_{+} are the same by Lemma 3.19, it follows by their finite dimensionality that the eigenspace structure of B^0_{+} is the same as the μ-eigenspace structure of B and hence, by Lemma 3.20, equal to the eigenspace structure of B'' modulo the span of the h_j's. ■

COROLLARY 3.21. *Aside from the points* $z = i\lambda_j$ *for the relevant* λ_j's, *the poles of* $\mathcal{S}'$ *in the upper half plane are at the points* $z = -i\mu$ *for the* μ*-eigenvalues of* B.

PROOF. According to the remark following Theorem 2.9, the scattering matrix $\mathcal{S}''$ will have poles at $-i$ times the eigenvalues of B''. According to Theorem 3.15 if we neglect the $-\lambda_j$'s for the relevant λ_j's, the eigenvalues of B'' coincide with the μ-eigenvalues of B. Finally the relation

$$\mathcal{S}' = \mathcal{S}_{+}^{-1}\mathcal{S}''\mathcal{S}_{-}^{-1}$$

shows that, aside from the zeros of $\mathcal{S}_{\pm}^{-1}$ which occur at $i\lambda_j$ for the relevant λ_j's, the poles of $\mathcal{S}'$ in the upper half plane coincide with those of $\mathcal{S}''$.

§4. THE LAPLACE-BELTRAMI OPERATOR FOR THE MODULAR GROUP

The modular group consists of fractional linear transformations

$$w \to \frac{aw+b}{cw+d}, \qquad ad - bc = 1, \tag{4.1}$$

where a, b, c and d are integers. A fundamental domain F for this group in the Poincaré plane Π, consists of the points x, y satisfying

$$-1/2 < x < 1/2, \quad 1 < x^2 + y^2, \quad x + iy = w; \tag{4.2}$$

F is bounded by two straight line segments and an arc of the unit circle and looks like this:

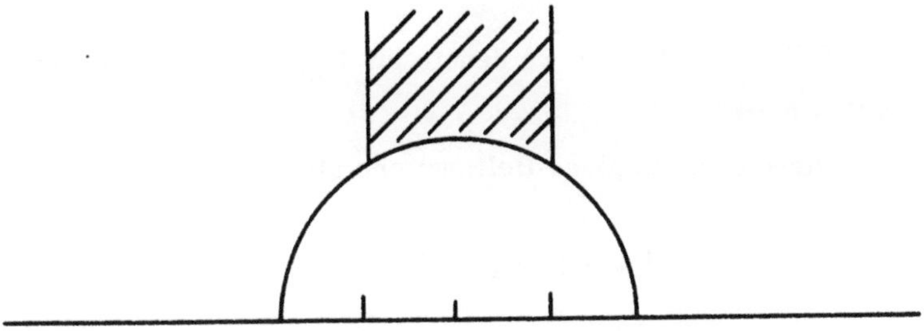

The transformation,

$$w \to w + 1 \tag{4.3}$$

carries the left segment bounding F into the right segment bounding F. The transformation

$$w \to -\frac{1}{w} \tag{4.4}$$

carries the left half of the circular arc bounding F into the right half.

With the above identification of its boundary points, the closure of F forms a manifold $\mathfrak{F}$. The points $\pm 1/2 + i\sqrt{3}/2$ and the point i are fixed points of some elements of the group (4.1) and need appropriate local parameters.

One can think of automorphic functions as functions on the manifold $\mathfrak{F}$ obtained by identifying boundary points of F, or one can think of them as functions defined on F whose boundary values satisfy certain matching conditions. For functions with continuous first derivatives these matching conditions take the following form:

$$(4.5) \qquad u(p) = u(\bar{p}), \qquad u_n(p) = -u_n(\bar{p})$$

for all pairs of boundary points p and $\bar{p}$ which have been identified with each other; here u_n denotes the outward normal derivative of u.

We introduce the L_2 scalar product with respect to the invariant metric (1.4):

$$(4.6) \qquad (f, g) = \iint_F \frac{f(x,y)\, g(x,y)}{y^2}\, dx\, dy \, .$$

We denote $(f, f)^{\frac{1}{2}}$ by $\|f\|$ and we denote by $L_2(F)$ the completion of $C_o(F)$ in this norm.

Let L_o denote the Laplace-Beltrami operator defined in (1.6):

$$(4.7) \qquad L_o = y^2(\partial_x^2 + \partial_y^2) = y^2\Delta \, .$$

L_o acts on automorphic functions; its precise domain is defined as follows:

To start with we take the domain of L_o to consist of all C^2 functions with compact support, which satisfy the boundary conditions (4.5). Integration by parts yields the following identity for u, v in the domain of L_o:

$$(4.8) \qquad (u, L_o v) = \iint_F u \Delta v \, dx\, dy = -\iint_F (u_x v_x + u_y v_y)\, dx\, dy \, ;$$

note that the boundary integrals are zero because of the boundary conditions (4.5).

We draw two conclusions from the relation (4.8):

a) L_o is symmetric

b) L_o is negative.

There is a well-known procedure, called the *Friedrichs extension*, for enlarging the domain of such a symmetric, densely defined semi-bounded operator so that the extended operator is *self-adjoint*. For sake of completeness we sketch the extension of L_o.

We define the bilinear form $C(u, v)$ by

(4.9) $$C(u,v) = (u,v) - (u, L_o v) .$$

Using (4.8) we get

(4.9)′ $$C(u,v) = \iint_F \left(u_x v_x + u_y v_y + \frac{uv}{y^2}\right) dx\,dy .$$

This shows that

a) $C(u, v)$ is symmetric

b) $C(u, u) \geq \|u\|^2$.

We denote by $\mathcal{C}$ the closure in the C norm of the domain of L_o; it follows from (4.9)′ that $\mathcal{C}$ is a subspace of L_2, consisting of all functions which are square integrable with respect to the invariant metric (1.4) and whose first derivatives are square integrable with respect to Lebesgue measure.

For any u in the domain of L_o set

(4.10) $$u - L_o u = f .$$

Taking the L_2 scalar product with any v in $\mathcal{C}$ and using the definition of C in (4.9) we get

(4.11) $$C(u,v) = (f,v) .$$

Now let f be any L_2 function; for any v in $\mathcal{C}$, we have

$$|(f,v)| \leq \|f\|\,\|v\| \leq \|f\|\,C^{\frac{1}{2}}(v)\ .$$

That is, (f,v) is a bounded linear functional of v in the C-norm. According to the Riesz representation theorem such a linear functional is the C-scalar product of v with an element u of $\mathcal{C}$, for which $C^{\frac{1}{2}}(u) \leq \|f\|$. For this u, relation (4.11) holds; we call this u a *weak solution* of (4.10), and *define the set of all such* u *to constitute the domain of the extended* L_o.

The extended L_o has the following properties:

a) its domain is contained in $\mathcal{C}$

b) L_o is symmetric

c) $\lambda = 1$ belongs to the resolvent set of L_o.

Property (a) is a direct consequence of the construction of u; property (b) follows from the symmetry of the bilinear form $C(u,v)$. Property (c) means that equation (4.10) has a solution u in the domain of the extended L_o, and that $\|u\| \leq \|f\|$. Since the extension was constructed so that (4.10) has a solution u which satisfies $C^{\frac{1}{2}}(u) \leq \|f\|$, and since $\|u\| \leq C^{\frac{1}{2}}(u)$, property (c) follows.

From now on the extended L_o will be called simply L_o.

Having shown that L_o is self-adjoint, we turn to its spectral theory:

THEOREM 4.1. a) *Every point of* $(-\infty, -1/4)$ *belongs to the spectrum of* L_o.

b) *Outside of the interval* $(-\infty, -1/4)$ *the spectrum of* L_o *consists of a finite number of eigenvalues each of finite multiplicity.*

REMARK. Part (a) of this result is relatively crude; the full story, which will appear as a byproduct of the scattering theory developed in Section 5, 6, and 7, is this:

a) The whole interval $(-\infty, -1/4)$ belongs to the continuous spectrum of L_o which has multiplicity one.

b) L_o has no singular spectrum but has infinitely many point eigenvalues imbedded in the continuous spectrum, accumulating at $-\infty$.

Part (a) of Theorem 4.1 was included as motivation of the setup in Section 5; part (b) however plays an essential role in the logical structure of our development.

PROOF. We shall use the following simple, general criterion, valid for any self-adjoint operator L:

If the interval $[\lambda-\varepsilon, \lambda+\varepsilon]$ is free of the spectrum of L, then

$$\|(L-\lambda)^{-1}\| < \frac{1}{\varepsilon} . \tag{4.12}$$

It follows conversely that if (4.12) fails to hold, then some point of $[\lambda-\varepsilon, \lambda+\varepsilon]$ belongs to the spectrum of L. Now (4.12) fails to hold if there is an f such that

$$\|(L-\lambda)^{-1}f\| \geq \frac{1}{\varepsilon}\|f\| .$$

Denoting $(L-\lambda)^{-1}f$ by u this inequality can be rewritten as follows:

$$\varepsilon\|u\| \geq \|(L-\lambda)u\| . \tag{4.13}$$

If for every positive ε there is a u_ε for which inequality (4.13) *is satisfied, then λ belongs to the spectrum of* L. The functions u_ε can be thought of as approximate eigenfunctions of L.

To apply this idea to the operator L_o, we start with a generalized eigenfunction, $y^{\frac{1}{2}+i\mu}$, μ real:

$$\begin{aligned} L_o y^{\frac{1}{2}+i\mu} &= y^2\partial_y^2 y^{\frac{1}{2}+i\mu} \\ &= (1/2+i\mu)(-1/2+i\mu)y^{\frac{1}{2}+i\mu} = -(1/4+\mu^2)y^{\frac{1}{2}+i\mu} = \lambda y^{\frac{1}{2}+i\mu} . \end{aligned} \tag{4.14}$$

This relation is purely formal; the function $y^{\frac{1}{2}+i\mu}$ does not belong to the domain of L_o because

a) it does not satisfy the boundary conditions (4.5) on the circular part of the boundary.

b) it does not belong to L_2.

We shall repair both defects by applying two C^∞ correction factors ϕ and ψ defined as follows:

(4.15) $$\phi(y) = \begin{cases} 0 & \text{for} \quad y < 2 \\ 1 & \text{for} \quad 3 < y \end{cases}$$

and

(4.16) $$\psi(\ell) = \begin{cases} 1 & \text{for} \quad \ell < 1 \\ 0 & \text{for} \quad 2 < \ell . \end{cases}$$

Now we set

(4.17) $$u_\varepsilon = \psi(\varepsilon \log y)\phi(y)\, y^{\frac{1}{2}+i\mu} .$$

Clearly, thanks to the factor ϕ, u_ε is zero for $y \le 1$ and so satisfies trivially the boundary condition (4.5) on the circular part of the boundary. The ψ factor makes the support of u_ε compact, so that u_ε is L_2. Consequently u_ε belongs to the domain of L_o as originally defined.

We now calculate $L_o u_\varepsilon$; using (4.14) we write what we get as

(4.18) $$L_o u_\varepsilon = \lambda u_\varepsilon + r_\varepsilon + s_\varepsilon ,$$

where r_ε denotes all terms which contain derivatives of ϕ and s_ε denotes all terms which contain derivatives of ψ but not of ϕ.

ASSERTION. $\|r_\varepsilon\|$ and $\|s_\varepsilon\|$ are uniformly bounded for all $\varepsilon < 1$;

(4.19) $$\|r_\varepsilon\| + \|s_\varepsilon\| \le \text{const}.$$

PROOF. It follows from (4.15) that the derivatives of ϕ are zero outside the y interval $(2, 3)$; from this the boundedness of $\|r_\varepsilon\|$ follows. To estimate s_ε we first write it out:

$$s_\varepsilon = \phi(1+2i\mu)\varepsilon\psi' y^{\frac{1}{2}+i\mu} + \phi\{\psi''\varepsilon^2 - \psi'\varepsilon\} y^{\frac{1}{2}+i\mu} .$$

Clearly

$$|s_\varepsilon(y)| \leq \text{const. } \varepsilon y^{\frac{1}{2}} ,$$

and furthermore the support of $s_\varepsilon(y)$ is contained in

$$1 < \varepsilon \log y < 2$$

i.e. in

$$\exp 1/\varepsilon < y < \exp 2/\varepsilon .$$

Therefore

$$\|s_\varepsilon\|^2 = \int \frac{|s_\varepsilon(y)|^2\, dy}{y^2} \leq \text{const. } \varepsilon^2 \int_{\exp 1/\varepsilon}^{\exp 2/\varepsilon} \frac{dy}{y} = \text{const. } \varepsilon .$$

This proves the uniform boundedness of $\|s_\varepsilon\|$.

Next we compute $\|u_\varepsilon\|$:

(4.20) $$\|u_\varepsilon\|^2 = \int \frac{|u_\varepsilon(y)|^2\, dy}{y^2} \geq \int_3^{\exp 1/\varepsilon} \frac{1}{y}\, dy = \frac{1}{\varepsilon} - \log 3 .$$

From (4.18) and (4.19) we obtain

$$\|(L_0 - \lambda)u_\varepsilon\| \leq \text{const} ;$$

combining this with (4.20) we get

$$\|(L_0 - \lambda)u_\varepsilon\| \leq \text{const. } \sqrt{\varepsilon}\, \|u_\varepsilon\| ,$$

which is inequality (4.13). Therefore we conclude that λ belongs to the spectrum of L_o.

According to (4.14),

$$\lambda = -(1/4 + \mu^2) ,$$

μ an arbitrary real number; the set of such λ fill out the interval $(-\infty, -1/4)$, therefore that whole interval belongs to the spectrum of L_o. This completes proof of part (a).

We turn now to part (b); we introduce the operator

$$L = L_o + 1/4 ; \tag{4.21}$$

part (b) asserts that the spectrum of L above zero consists of a finite number of eigenvalues, each of finite multiplicity.

Using (4.8) we have the following expression for the quadratic form associated with L:

$$-(u, Lu) = \iint_F \left(u_x^2 + u_y^2 - \frac{u^2}{4y^2}\right) dx\, dy . \tag{4.22}$$

Let a be any number ≥ 2; we divide F into two parts, one above and the other below $y = a$:

$$F_0 = F \cap \{y < a\}, \quad F_1 = F \cap \{y \geq a\} . \tag{4.23}$$

We claim that the following identity holds:

$$\iint_{F_1} \left(u_y^2 - \frac{u^2}{4y^2}\right) dx\, dy = \iint_{F_1} y\left[\partial_y\left(\frac{u}{\sqrt{y}}\right)\right]^2 dx\, dy - \int_{y=a} \frac{u^2}{2y}\, dx . \tag{4.24}$$

To prove this identity we merely expand the integral over F_1 on the right side as

$$\iint_{F_1} \left(u_y^2 - \frac{u_y u}{y} + \frac{u^2}{4y^2}\right) dx\, dy$$

and integrate by parts the middle term. Combining (4.22) and (4.24) we get

$$(4.25)\qquad -(u, Lu) = \iint\limits_{F_1} \left(u_x^2 + y\left[\partial_y\left(\frac{u}{\sqrt{y}}\right)\right]^2\right) dx\,dy$$

$$+ \iint\limits_{F_0} \left(u_x^2 + u_y^2 - \frac{u^2}{4y^2}\right) dx\,dy - \int\limits_{y=a} \frac{u^2}{2y}\, dx\ .$$

We shall estimate now the line integral appearing in (4.25):

LEMMA 4.2. *For* $a > 2$

$$(4.26)\qquad \int\limits_{y=a} \frac{u^2}{2y}\, dx \le \iint\limits_{F_0} \left(3/2\, \frac{u^2}{y^2} + 1/2\, u_y^2\right) dx\,dy\ .$$

PROOF. We introduce the auxiliary function

$$\phi(y) = \frac{2y - a}{a}\ ,$$

equal to 1 at $y = a$ and equal to 0 at $y = a/2$. We write

$$(4.27)\qquad \frac{u^2(x,a)}{2a} = \frac{1}{2a} \int_{a/2}^{a} \partial_y(\phi u^2)\, dy\ .$$

The integral on the right is

$$\int_{a/2}^{a} \left(\frac{\phi_y}{2a}\, u^2 + \frac{\phi}{a}\, uu_y\right) dy\ .$$

Using the fact that $\phi_y = 2/a$ for $a/2 < y < a$, we see that the first integral is less than

$$\frac{1}{a^2}\int_{a/2}^{a} u^2\,dy \le \int_{a/2}^{a} \frac{u^2}{y^2}\,dy \;. \tag{4.28}$$

We estimate the second integral by the Schwarz inequality; using the facts $|\phi| \le 1$, $y < a$, and applying the arithmetic geometric mean inequality we get

$$\int_{a/2}^{a} \frac{\phi}{a}\, u u_y\,dy \le \left\{\int_{a/2}^{a} \frac{u^2}{y^2}\,dy \int_{a/2}^{a} u_y^2\,dy\right\}^{\frac{1}{2}} \tag{4.28$'$}$$

$$\le 1/2 \int_{a/2}^{a} \left(\frac{u^2}{y^2} + u_y^2\right) dy \;.$$

Adding (4.28) and (4.28)′ we see that (4.27) is bounded by

$$\int_{a/2}^{a} \left(3/2\,\frac{u^2}{y^2} + 1/2\,u_y^2\right) dy \;.$$

Integrating this with respect to x we get (4.26). ∎

Next combining (4.25) and (4.26) we get

$$-(u, Lu) \ge \iint_{F_1} \left(u_x^2 + y\left[\partial_y\left(\frac{u}{\sqrt{y}}\right)\right]^2\right) dx\,dy$$

$$+ \iint_{F_0} \left(u_x^2 + 1/2\,u_y^2 - \frac{7}{4}\,\frac{u^2}{y^2}\right) dx\,dy \;.$$

We now introduce the quadratic form

$$K(u) = \iint_{F_0} \frac{u^2}{y^2}\,dx\,dy \;. \tag{4.29}$$

In terms of K the above inequality can be rewritten as follows:

$$(4.30)\qquad 2K(u) - (u, Lu) \geq \iint_{F_1} \left(u_x^2 + y\left[\partial_y\left(\frac{u}{\sqrt{y}}\right)\right]^2\right) dx\, dy$$

$$+ \iint_{F_0} \left(u_x^2 + 1/2\, u_y^2 + 1/4\, \frac{u^2}{y^2}\right) dx\, dy\ .$$

According to a classical result of *Rellich*, the quadratic form consisting of the square integral of a function u over a compact domain F_0 with decent boundary is compact with respect to the quadratic form consisting of the square integral of the function and its first derivatives over F_0. From this we deduce the following:

LEMMA 4.3. *The quadratic form* K(u) *is compact with respect to the quadratic form*

$$C_0(u) = \iint_{F_0} \left(u_x^2 + u_y^2 + \frac{u^2}{y^2}\right) dx\, dy\ .$$

Compactness implies that for any positive ε

$$(4.31)\qquad K(u) < \varepsilon\, C_0(u)$$

on some subspace of finite codimension. Now take ε to be any number $< 1/8$; since the right side of inequality (4.30) is $\geq \frac{1}{4} C_0$ we deduce from (4.30) and (4.31) that

$$(u, Lu) < 0$$

on a subspace of finite codimension. It follows from this that the subspace on which L is non-negative is finite dimensional. This completes the proof of part (b) of Theorem 4.1. ■

We conclude this section with some additional information about L which will be needed in Section 6.

I) The definition of L_0 makes it clear that its domain consists of functions whose first derivatives are square integrable. Actually more is true:

THEOREM 4.4. *Functions in the domain of* L_0 *have locally square integrable second derivatives. Furthermore the* L_2 *norm of* u_{xx}, u_{xy}, u_{yy} *over any compact set is bounded by a constant times the* L_2 *norm of* u *and* $L_0 u$ *over any larger set.*

PROOF. We can deduce this estimate from classical results concerning the Laplace operator, except for the presence of possible singularities at the three vertices of F, that is at the points $\pm 1/2 + i\frac{\sqrt{3}}{2}$ and i. All other boundary points become regular interior points after identification of corresponding boundary points. ■

In an appendix to this section we shall show that the possible singularities at the three exceptional points are removable.

II) For the modular group, the fundamental domain F has bilateral symmetry around the y-axis. Reflection across the y axis leaves the Laplace-Beltrami operator invariant; it follows that the eigenspaces of reflection, the even and odd functions, reduce the operator L. The pairs of boundary points p and $\overline{p}$ which are identified with each other are images of each other under reflection. Consider now the boundary conditions (4.5):

$$u(p) = u(\overline{p}), \qquad u_n(p) = -u_n(\overline{p}) .$$

For an odd function,

$$u(p) = -u(\overline{p}) ;$$

therefore in order to satisfy the first boundary condition we must have

$$u(p) = 0 \ . \tag{4.32}$$

On the other hand the second boundary condition is automatically satisfied.

It is just the other way for even functions; there the first boundary condition is automatically satisfied, while the second condition demands that

$$u_n(p) = 0 \ . \tag{4.33}$$

So for odd functions the automorphic boundary conditions amount to the *Dirichlet* boundary condition (4.32), while for even functions they become the *Neumann* boundary condition (4.33).

Appendix to Section 4. Regularity near a vertex.

We wish to establish the regularity of a function u in the domain of L_o near a vertex of the fundamental domain F; in the case of the modular group this would be near the points $\pm 1/2 + i\frac{\sqrt{3}}{2}$ and i. Only a finite number of translates of the domain, that is γF's, contain such a vertex; and by extending u by means of these γ's into the punctured disc about such a vertex the so-extended u satisfies a relation of the form

$$\Delta u = g \ ,$$

where u and its first derivatives are square integrable and g is square integrable. It is convenient at this point to translate the vertex into the origin.

Next denote by v the solution of

$$\Delta v = g$$

in the entire disc where $v = u$ on the boundary of the disc. By elliptic theory, the L_2 norms of v and its derivatives of order ≤ 2 on any concentric subdisc are bounded by a constant times the L_2 norm of v and g on the disc. It therefore suffices to prove that $u = v$. To this end we set

$$w = u - v \ ;$$

then w is harmonic in the punctured disc, vanishes on the boundary of the disc and w, ∇w are square integrable in the disc.

In analogy with the Laurent series we can express w as the sum of a function harmonic in the disc, a $\log 1/r$ term and a convergent series in powers of $1/r$:

(4.34) $$w(r,\Theta) = a_0 \log \frac{1}{r} + a_1 \frac{\cos\Theta}{r} + b_1 \frac{\sin\Theta}{r} + \cdots + \text{harmonic functions}.$$

It follows from this that

(4.35) $$a_n = \lim_{r\to 0} \frac{r^n}{\pi} \int_0^{2\pi} w(r,\Theta)\cos n\Theta\, d\Theta,\ b_n = \lim_{r\to 0} \frac{r^n}{\pi} \int_0^{2\pi} w(r,\Theta)\sin n\Theta\, d\Theta.$$

On the other hand

$$w(r,\Theta) = -\int_r^{r_o} \frac{\partial w}{\partial r}\, dr + w(r_o,\Theta)\ ;$$

and hence

$$|w(r,\Theta)|^2 \le 2\left|\int_r^{r_o} \frac{\partial w}{\partial r}\, dr\right|^2 + 2|w(r_o\Theta)|^2\ .$$

Applying the Schwarz inequality to the integral on the right:

$$\left|\int_r^{r_o} \frac{\partial w}{\partial r}\, dr\right|^2 \le \int_r^{r_o} \left|\frac{\partial w}{\partial r}\right|^2 r\, dr \cdot \log \frac{r_o}{r}$$

and integrating with respect to Θ, we get

(4.36) $$\int_0^{2\pi} |w(r,\Theta)|^2\, d\Theta = O\left(\log \frac{1}{r}\right)$$

as $r \to 0$. We conclude from (4.35) and (4.36) that $a_n = 0 = b_n$ for all

$n > 0$. This means that w has at most logarithmic singularity at the origin, but even this is excluded by the square integrability of $\frac{\partial w}{\partial r}$.

It follows that w has a removable singularity at the origin; that is w is harmonic in the disc. Since w equals zero on the boundary of the disc, we conclude by the maximum principle that it is identically zero in the disc. Retracing our steps, we see that u has the stated regularity properties near a vertex.

§5. THE AUTOMORPHIC WAVE EQUATIONS

We are now ready to begin our study of the automorphic wave equation:

$$u_{tt} = y^2 \Delta u + \frac{1}{4} u \; . \tag{5.1}$$

In order to motivate the presence of $\frac{1}{4}u$ in (5.1), we recall that the modified scattering theory developed in Section 3 is applicable to a second order equations of the form

$$u_{tt} = Lu \tag{5.2}$$

only if

a) the continuous spectrum of L consists of $(-\infty, 0)$ and is of uniform multiplicity;

b) L has only a finite number of positive eigenvalues.

Now according to Theorem 4.1, the operator $L = L_o + 1/4$, where L_o is the automorphic Laplace-Beltrami operator (4.7), has property (b) and very likely property (a).

In Section 6 we shall show that the theory of Section 3 is indeed applicable to the automorphic wave equation. In the present section we elaborate on the abstract theory for second order equations developed in the first part of Section 3.

The energy associated with equation (5.1) is given by (3.5) as

$$E = \|u_t\|^2 - (u, Lu) \; . \tag{5.3}$$

Using (4.25) we can rewrite this expression as

$$(5.3)' \qquad E = \iint_{F_o} \left(u_x^2 + u_y^2 - \frac{u^2}{4y^2} + \frac{u_t^2}{y^2}\right) dx\,dy$$

$$+ \iint_{F_1} \left(u_x^2 + y\left[\partial_y\left(\frac{u}{\sqrt{y}}\right)\right]^2 + \frac{u_t^2}{y^2}\right) dx\,dy - \int_{y=a} \frac{u^2}{2y}\,dx\ .$$

The energy form is not positive on $\mathcal{H}_o$, defined in Section 3 as consisting of all data $f = \{f_1, f_2\}$ with f_1 in the domain of $|L|^{\frac{1}{2}}$ and f_2 in $L_2(F)$. However E is non-negative on the subspace $\mathcal{H}_o'$ of $\mathcal{H}_o$ for which both f_1 and f_2 are orthogonal in the L_2 innerproduct to the eigenfunctions of L having positive eigenvalues. We have shown in Section 4 that there are only a finite number of such eigenfunctions. Equivalently $\mathcal{H}_o'$ can be defined in the notation of Section 3 as the E-orthogonal complement in $\mathcal{H}_o$ of $\mathcal{P}$, where $\mathcal{P}$ is spanned by the eigenfunctions $\{f_j^{\pm}\}$ of A defined in (3.12).

On $\mathcal{H}_o'$ the energy is non-negative, which makes it possible to complete $\mathcal{H}_o'$ in the energy norm; this completion was denoted by $\mathcal{H}_E'$ in Section 3.

Because of its indefinite character, the energy form is difficult to handle technically, even on subspaces such as $\mathcal{H}_E'$ on which E is non-negative. For this reason we have been led to introduce a new quadratic form G, obtained by adding a relatively small quantity to E. The resulting form is not only positive but obviously so. The rest of this section is devoted to defining and studying this new norm, especially its relation to the energy norm and the operators $U(t)$.

The quantity which we add to E is suggested by inequality (4.30):

$$(5.4) \qquad 2K(u) - (u, Lu) \geq \iint_{F_1} \left(u_x^2 + y\left[\partial_y\left(\frac{u}{\sqrt{y}}\right)\right]^2\right) dx\,dy$$

$$+ \iint_{F_o} \left(u_x^2 + \frac{1}{2}\,u_y^2 + \frac{1}{4}\,\frac{u^2}{y^2}\right) dx\,dy\ .$$

We define

$$K(u) = K(u_1) \equiv \iint_{F_o} \frac{u_1^2}{y^2}\, dx\, dy \tag{5.5}$$

and

$$G(u) = E(u) + 2K(u) . \tag{5.6}$$

The properties of G that will be used later on are summarized in

LEMMA 5.1. a) *The quadratic form*

$$G'(u) = \iint_{F_o} \left(u_{1x}^2 + u_{1y}^2 + \frac{u_1^2}{y^2} + \frac{u_2^2}{y^2} \right) dx\, dy \tag{5.7}$$

$$+ \iint_{F_1} \left(u_{1x}^2 + y \left[\partial_y \left(\frac{u_1}{\sqrt{y}} \right) \right]^2 + \frac{u_2^2}{y^2} \right) dx\, dy$$

is equivalent with G(u), i.e. *for all* u

$$kG'(u) \le G(u) \le KG'(u) \tag{5.7'}$$

with some constants k, K.

b) *For any compact subset* C *of* F

$$\iint_C u_1^2\, dx\, dy \le \text{const. } G(u) . \tag{5.7''}$$

PROOF. The first part of inequality (5.7)′ of part (a) is just a restatement of (5.4), after making use of (5.3). To prove the second part of inequality (5.7)′ we combine (5.3)′ and (5.5). To prove part (b), we represent u_1 as follows:

$$\frac{u_1(x,y)}{\sqrt{y}} = \frac{u_1(x,a)}{\sqrt{a}} + \int_a^y \partial_y\left(\frac{u_1}{\sqrt{y}}\right) dy \;.$$

From this relation we derive the estimate

$$\iint_a^b u_1^2(x,y)\,dx\,dy \le \text{const.}\left\{\int u_1^2(x,a)\,dx + \iint_a^b y\left[\partial_y\left(\frac{u_1}{\sqrt{y}}\right)\right]^2 dx\,dy\right\}.$$

Combining this with the inequalities (4.26) and (5.7) yields (5.7)″. ∎

We now denote by $\mathcal{H}_G$ and $\mathcal{H}'_G$ the completion in the G-norm of the subspaces $\mathcal{H}_o$ and $\mathcal{H}'_o$, respectively, defined following (3.18). $\mathcal{H}'_G$ is a subspace of $\mathcal{H}_G$ which is clearly related to $\mathcal{H}'_E$.

Since the G-norm is larger than the E-norm on $\mathcal{H}'_o$, every Cauchy sequence in the sense of the G-norm is also a Cauchy sequence in the E-norm; this results in a *natural injection* of the G-completion $\mathcal{H}'_G$ into the E-completion $\mathcal{H}'_E$ which is obviously continuous.

THEOREM 5.2. *The natural injection map*

(5.8) $$\mathcal{H}'_G \to \mathcal{H}'_E$$

defined above is onto and its kernel is finite dimensional.

PROOF. Denote the kernel of this mapping by $\mathcal{J}$; it consists of all functions in $\mathcal{H}'_G$ with zero energy. It follows then from (5.6) that for f in $\mathcal{J}$

(5.9) $$G(f) = 2K(f) \;.$$

According to Lemma 4.3, $K(f) = K(f_1)$ is compact with respect to $C_o(f_1)$ as defined there. Since $G(f) \ge \frac{1}{4} C_o(f_1)$, it follows that $K(f_1)$ is compact with respect to $G(f)$. This implies that the unit ball in $\mathcal{J}$ with respect to the G-norm is compact in the K-norm. But by (5.9) this unit ball is also the unit ball in the 2K-norm. According to a classical observation of F. Riesz, unit balls are compact only in finite dimensional spaces – this proves that $\mathcal{J}$ is finite dimensional.

Next take f in $\mathcal{H}'_0$ and denote by j the G-orthogonal projection of f into $\mathcal{J}$. We now show that

$$g = f - j \quad \text{in} \quad \mathcal{H}'_G$$

maps into f in $\mathcal{H}'_E$ under the natural injection. The elements of $\mathcal{J}$ are limits in the G-norm of elements of $\mathcal{H}'_0$ which form a null sequence in the E-norm; that is, there exist elements j'_n in $\mathcal{H}'_0$ such that

$$G(j - j'_n) \to 0 \tag{5.10}$$

and

$$E(j'_n) \to 0 \ . \tag{5.10$'$}$$

The sequence

$$g'_n = f - j'_n \quad \text{in} \quad \mathcal{H}'_0$$

forms a Cauchy sequence in both norms, converging to g in the G-norm because of (5.10) and, because of (5.10)′, to f in the E-norm. This proves that g maps into f under the natural injection.

In order to prove that the mapping is onto, let f be any element in $\mathcal{H}'_E$ and choose a Cauchy sequence of elements $\{f_n\}$ in $\mathcal{H}'_0$ which tend to f in the E-norm. Denote the G-orthogonal projection of f_n into $\mathcal{J}$ by j_n and set

$$g_n = f_n - j_n \ .$$

As above, g_n belongs to $\mathcal{H}'_G$ and maps onto f_n in $\mathcal{H}'_E$ under the natural injection. If it can be shown that the $\{g_n\}$, or some subsequence thereof, converges to a limit $g \in \mathcal{H}'_G$ in the G-norm, then since g_n maps into f_n and the f_n converge to f in the E-norm, it will follow that g maps into f.

We first show that the $G(g_n)$ are bounded. For assume not; then for some subsequence

$$a_n^2 = G(g_n)$$

tends to infinity. Define

$$h_n = g_n/a_n \ ;$$

then

$$G(h_n) = 1 \ ; \tag{5.11}$$

and since the $E(f_n)$ are bounded,

$$E(h_n) = \frac{E(g_n)}{a_n^2} = \frac{E(f_n)}{a_n^2} \leq \frac{\text{const.}}{a_n^2} \tag{5.12}$$

tends to zero. By compactness (Lemma 4.3) we can select a subsequence of the h_n which converge in the K-norm. Since by (5.12) the $\{h_n\}$ is a null sequence in the E-norm, it follows from the definition (5.6) of G that $\{h_n\}$ is also a Cauchy sequence in the G-norm. Denote its limit by h. By (5.12) we see that $E(h) = 0$ so that h belongs to $\mathcal{J}$. On the other hand by choice h is G-orthogonal to $\mathcal{J}$; this implies that $h = 0$. But we deduce from (5.11) that $G(h) = 1$, a contradiction, into which we got by assuming $G(g_n)$ to be unbounded. So we conclude that $G(g_n)$ is in fact bounded.

It now follows by compactness from the boundedness of $G(g_n)$ that some subsequence of $\{g_n\}$ is a Cauchy sequence in the K-norm. Since $\{f_n\}$ is already a Cauchy sequence in the E-norm, we conclude from (5.6) that this subsequence is a Cauchy sequence in the G-norm. As we have already remarked, the limit of this Cauchy sequence in $\mathcal{H}'_G$ is mapped into f under the natural injection (5.8). This completes the proof of Theorem 5.2. ∎

COROLLARY 5.3. *The natural mapping*

$$\mathcal{H}'_G/\mathcal{I} \to \mathcal{H}'_E$$

is a homeomorphism.

PROOF. It follows from Theorem 5.2 that the mapping is onto; clearly it is one-to-one. Since $E(f) \leq G(f)$, the mapping is norm decreasing; since both spaces are Hilbert spaces, it follows then from the closed graph theorem that the mapping is a homeomorphism. ∎

REMARK 1. One can avoid reference to the closed graph theorem and prove the boundedness of the inverse directly using the compactness of K.

REMARK 2. Corollary 5.3 shows that the elements of $\mathcal{H}'_E$ are not functions but equivalence classes of functions; the reason for this is that $\mathcal{H}'_E$ is the completion of a function space in the energy norm, which contains square integrals of first derivatives but not of the function itself. The following simple example elucidates what has happened:

Consider all functions $u(y)$ with compact support on the positive real axis, and introduce as energy norm

$$E(u) = \int_0^\infty u_y^2 \, dy \ .$$

The function $u \equiv 1$ is the limit in this norm of a sequence of functions $\{u_n\}$ with compact support:

$$u_n = \begin{cases} 1 & \text{for} \quad y < n \\ 2 - \frac{y}{n} & \quad\quad n < y < 2n \\ 0 & \text{for} \quad 2n < y \ . \end{cases} \tag{5.13}$$

Therefore elements of the completion are functions modulo arbitrary constants.

REMARK 3. The set $\mathcal{J}$ can also be characterized as the degenerate set for the E-form in $\mathcal{H}_G$; that is as the set of all j for which

$$E(j, \mathcal{H}_G) = 0 \ . \tag{5.14}$$

For if j satisfies (5.14), then in particular j is E-orthogonal to $\mathcal{P}$, hence in $\mathcal{H}'_G$, and since in addition it has zero energy, it belongs to $\mathcal{J}$. Conversely if j belongs to $\mathcal{J}$ one can reverse this argument by noting that since E is non-negative on $\mathcal{H}'_G$, any j in $\mathcal{J}$ will be E-orthogonal to all of $\mathcal{H}'_G$ by the Schwarz inequality. Finally, since $\mathcal{H}'_G$ is E-orthogonal to $\mathcal{P}$, j will be E-orthogonal to all of $\mathcal{H}_G$.

We turn now to the study of A on $\mathcal{H}_G$ with the assurance that we are now working with an honest function space.

THEOREM 5.4. a) *The operator*

$$A = \begin{pmatrix} O & I \\ L & O \end{pmatrix} , \tag{5.15}$$

its domain suitably defined in $\mathcal{H}_G$, *generates a group of operators* U(t) *which grows at most exponentially.*

b) $\mathcal{J}$ *consists of the nullspace of* A.

c) *This definition of* U(t) *on* $\mathcal{H}_G$ *is consistent with the definition of* U(t) *already given in* $\mathcal{H}'_E$. *That is the following diagram commutes:*

$$\begin{array}{ccc} \mathcal{H}'_G & \xrightarrow{U} & \mathcal{H}'_G \\ \downarrow & & \downarrow \\ \mathcal{H}'_E & \xrightarrow{U} & \mathcal{H}'_E \end{array} ,$$

where each vertical arrows is the natural injection (5.7) *of* $\mathcal{H}'_G$ *onto* $\mathcal{H}'_E$.

PROOF. We recall from Section 4 that the domain of L is a subspace of $\mathcal{C}$ and that $\mathcal{C}$ consists of functions u for which

$$C(u) = \iint_F \left(u_x^2 + u_y^2 + \frac{u^2}{y^2}\right) dx\,dy < \infty . \tag{5.16}$$

Moreover the domain of $|L|^{\frac{1}{2}}$ is precisely $\mathcal{C}$. We define the core domain of A to consist of the set of all data $f = \{f_1, f_2\}$ for which f_1 belongs to the domain of L and f_2 belongs to the domain of $|L|^{\frac{1}{2}}$. By definition (5.15),

$$Af = \begin{pmatrix} O & I \\ L & O \end{pmatrix} \begin{pmatrix} f_1 \\ f_2 \end{pmatrix} = \begin{pmatrix} f_2 \\ Lf_1 \end{pmatrix} ;$$

it follows from inequality (5.16) and the fact that Lf_1 is in $L_2(F)$ that Af belongs to $\mathcal{H}_G$. Finally we extend the operator A by *closure* in $\mathcal{H}_G$.

The following estimate will be useful:

LEMMA 5.5. *For all* f *in the domain of* A

$$|G(Af, f)| \leq 4G(f) . \tag{5.17}$$

PROOF. It suffices to prove (5.17) for all f in the core domain of A. By definition (5.6),

$$G(Af, f) = E(Af, f) + 2K(Af, f) .$$

We claim that

$$E(Af, f) = 0 ;$$

for by definition (5.3),

$$E(g, f) = -(g_1, Lf_1) + (g_2, f_2) ;$$

letting $\{g_1, g_2\} = Af = (f_2, Lf_1)$ we have

$$E(Af, f) = -(f_2, Lf_1) + (Lf_1, f_2) = 0 \tag{5.18}$$

on account of the symmetry of L. By definition (5.5) of K

$$K(g, f) = \iint_{F_o} \frac{g_1 f_1}{y^2} \, dx\, dy \ ;$$

and hence

$$|K(Af, f)| = \left| \iint_{F_o} \frac{f_2 f_1}{y^2} \, dx\, dy \right| \leq \frac{1}{2} \iint_{F_o} \frac{f_2^2 + f_1^2}{y^2} \, dx\, dy \ . \tag{5.19}$$

Using definition (5.6) of G and inequality (5.4), it follows from (5.19) that

$$|K(Af, f)| \leq 2G(f) \ .$$

Combining this with (5.18) we get (5.17).

LEMMA 5.6. *The operator* A *satisfies the Hille-Yosida criterion*:

$$\|(\lambda - A)^{-1}\|_G \leq \frac{1}{|\lambda| - 4} \quad \textit{for} \quad |\lambda| > 4 \ , \tag{5.20}$$

where the subscript G *denotes the operator norm with respect to the* G-*form*.

PROOF. We show first that the inequality

$$G^{\frac{1}{2}}((\lambda - A)^{-1} f) \leq \frac{1}{|\lambda| - 4} G^{\frac{1}{2}}(f) \tag{5.21}$$

holds for all f in the range of $\lambda - A$. Let

$$(\lambda - A)^{-1} f = u \ ,$$

so that

$$(\lambda - A) u = f \ . \tag{5.22}$$

Take the G-scalar product of both sides with u:

$$\lambda G(u) - G(Au, u) = G(f, u) .$$

Thus

$$|\lambda| G(u) \leq |G(Au, u)| + |G(f, u)| .$$

We estimate the first term on the right by (5.17) and the second by the Schwarz inequality. This gives

$$|\lambda| G(u) \leq 4G(u) + G^{\frac{1}{2}}(u) G^{\frac{1}{2}}(f)$$

which implies

$$G^{\frac{1}{2}}(u) \leq \frac{1}{|\lambda| - 4} G^{\frac{1}{2}}(f) .$$

Since $u = (\lambda - A)^{-1} f$ this is the desired inequality (5.21).

Next we show that the range of $(\lambda - A)$ is all of $\mathcal{H}_G$. For this purpose we write (5.22) in component form:

$$\begin{aligned} \lambda u_1 - u_2 &= f_1 \\ \lambda u_2 - L u_1 &= f_2 . \end{aligned} \tag{5.23}$$

Multiply the first relation in (5.23) by λ and add it to the second; this gives

$$(-L + \lambda^2) u_1 = \lambda f_1 + f_2 . \tag{5.24}$$

Suppose f_1 belongs to the domain of $|L|^{\frac{1}{2}}$ and f_2 belongs to L_2, then the right side of (5.24) belongs to L_2; since $-L_o = -L + 1/4$ is positive, it follows that for $\lambda^2 > 1/4$ equation (5.24) has a solution u_1 which belongs to the domain of L. The first of equations (5.23) shows that $u_2 = \lambda u_1 - f_1$ also belongs to the domain of $|L|^{\frac{1}{2}}$. Thus u_1 belongs to the domain of L and u_2 belong to the domain of $|L|^{\frac{1}{2}}$, which puts $u = \{u_1, u_2\}$ in the core domain of A. This shows that $f = \{f_1, f_2\}$

belongs to the range of $\lambda - A$ if f_1 belongs to the domain of $|L|^{\frac{1}{2}}$ and f_2 is in L_2. The set of such f is dense in $\mathcal{H}_G$. It follows then from inequality (5.21) and the fact that A is a closed operator that the range of $\lambda - A$ is all of $\mathcal{H}_G$. This completes the proof of Lemma 5.6. ∎

Because of Lemma 5.6, the Hille-Yosida theorem applies and A generates a group U(t) on $\mathcal{H}_G$. This completes the proof of part (a) of Theorem 5.4.

REMARK. It follows by polarity from (5.18) that A is *skew symmetric with respect to* E.

We turn now to part (b). Suppose first that $Af = 0$. We shall show that f belongs to $\mathcal{J}$, that is that f belongs to $\mathcal{H}'_G$ and that $E(f) = 0$. Using the E-skew symmetry of A we get

$$\pm \lambda_j E(f, f_j^{\pm}) = E(f, Af_j^{\pm}) = -E(Af, f_j^{\pm}) = 0 ;$$

this proves that f is E-orthogonal to $\mathcal{P}$ and hence belongs to $\mathcal{H}'_G$. Further it is clear from (5.15) that $f_2 = 0$ and since

$$0 = E(Af, g) = -E(f, Ag)$$

it follows that E(f) will vanish if we can approximate f_1 by the first component of Ag, that is by g_2. But this can obviously be done because g_2 can be any element of the domain of $|L|^{\frac{1}{2}}$ and $D(|L|^{\frac{1}{2}})$ fills out the first component set of $\mathcal{H}_o$, which is dense in $\mathcal{H}_G$.

To prove the converse assertion, let $j = \{j_1, j_2\}$ be an element of $\mathcal{J}$. It follows from (5.14) that j is E-orthogonal to $\mathcal{H}_G$ so that

$$0 = E(j, f) = -(j_1, Lf_1) + (j_2, f_2)$$

for all f in $\mathcal{H}_o$. Taking $f_1 = 0$ we conclude that $j_2 = 0$. It therefore follows that j_1 satisfies

$$Lj_1 = 0 \tag{5.25}$$

in the weak sense. Since L is an elliptic operator it follows as shown in the appendix to Section 4 that j_1 is smooth, satisfying the equation (5.25) pointwise as well as the boundary conditions; so we have, at least formally

$$Aj = \{j_2, Lj_1\} = 0 .$$

It remains to show that j actually belongs to the domain of A. For this we use the fact that j has finite G-norm; by (5.4) this implies that

$$\iint_{F_1} y\left[\partial_y\left(\frac{j_1}{\sqrt{y}}\right)\right]^2 dx\,dy < \infty . \tag{5.26}$$

Now let ψ be the auxiliary function introduced in Section 4:

$$\psi(\ell) = \begin{cases} 1 & \text{for} \quad \ell < 1 \\ 0 & \text{for} \quad \ell > 2 . \end{cases}$$

We set

$$j^{(n)} = \psi\left(\frac{\log y}{n}\right) j .$$

Then $j^{(n)}$ belongs to the domain of A; and it is easy to show that as $n \to \infty$, $j^{(n)} \to j$ in the G-norm.

We claim that $Aj^{(n)}$ is a null sequence in the G-norm.

We have already shown that $j_2 = 0$; so

$$Aj^{(n)} = A\psi j = \{0, L\psi j_1\} ,$$

and

$$G(Aj^{(n)}) = \iint_F \frac{(L\psi j_1)^2}{y^2}\, dx\,dy .$$

Using the fact that $Lj_1 = 0$ we have

(5.27)
$$Lj_1^{(n)} = L\psi j_1 = 2\psi' \frac{y}{n} \partial_y j_1 + \left[\frac{\psi''}{n^2} - \frac{\psi'}{n}\right] j_1$$

$$= \frac{2y^{3/2}}{n} \psi' \partial_y\left(\frac{j_1}{\sqrt{y}}\right) + \frac{\psi''}{n^2} j_1 = r_1 + r_2 \ .$$

Now

$$\|r_1\|^2 = \iint \frac{r_1^2}{y^2} dx\, dy = \frac{4}{n^2} \iint \psi'^2 y \left[\partial_y\left(\frac{j_1}{\sqrt{y}}\right)\right]^2 dx\, dy \ ;$$

since ψ' is bounded and the integral (5.26) is bounded, we deduce that

(5.28)
$$\|r_1\|^2 \leq \frac{\text{const.}}{n^2}$$

To estimate r_2 we use the representation

$$\frac{j_1(y)}{\sqrt{y}} - \frac{j_1(y_o)}{\sqrt{y_o}} = \int_{y_o}^{y} \partial_y\left(\frac{j_1}{\sqrt{y}}\right) dy \ .$$

We get, by the Schwarz inequality, that

(5.29)
$$\int \frac{j_1^2(x,y)}{y^2} dx \leq \frac{B_o}{y} + C_o \frac{\log y}{y} \ ,$$

where

$$B_o = \int \frac{2j_1^2(x, y_o)}{y_o} dx \ ,$$

$$C_o = \iint_{y > y_o} y \left[\partial_y\left(\frac{j_1}{\sqrt{y}}\right)\right]^2 dx\, dy \ .$$

It follows from (5.29) that

(5.30)
$$\|r_2\|^2 = \frac{1}{n^4} \int_{\exp n}^{\exp 2n} \psi''^2 \frac{j_1^2}{y^2}\, dx\, dy$$

$$\leq \frac{\text{const.}}{n^4} [B_o n + 3C_o n^2] \,.$$

Using the estimates (5.28) and (5.30) we get from (5.27) that

$$\|Lj_1^{(n)}\| \leq \|r_1\| + \|r_2\| \leq \frac{\text{const.}}{n},$$

which tends to 0 as $n \to \infty$. This completes the proof of part (b) of Theorem 5.4.

Part (c) of Theorem 5.4 is now easy to prove. First of all, $\mathcal{H}'_G$ is an invariant subspace for $U(t)$ and for A. Secondly $A\mathcal{J} = 0$ so $U(t)$ takes $\mathcal{J}$ into itself. It follows that A and $U(t)$ can be defined as operators on $\mathcal{H}'_G/\mathcal{J}$. The natural injection (5.8) which, according to Corollary 5.3, carries $\mathcal{H}'_G/\mathcal{J}$ onto $\mathcal{H}'_E$, carries $U(t)$ into a group of operators on $\mathcal{H}'_E$; we denote this group by U_G. The generator A_G of U_G contains the closure of A restricted to data $f = \{f_1, f_2\}$ in $\mathcal{H}'_o$ with f_1 in the domain of L and f_2 in the domain of $|L|^{\frac{1}{2}}$.

On the other hand in Section 3 we defined a unitary group on $\mathcal{H}'_E$; let us for the moment denote the generator of this group by A_E. We showed in Section 3 that A_E is the closure (in the E-norm) of A defined on the core domain consisting as above of data $f = \{f_1, f_2\}$ in $\mathcal{H}'_o$ with f_1 in the domain of L, f_2 in the domain of $|L|^{\frac{1}{2}}$. It follows that A_G is an extension of A_E; since both operators generate groups which grow at most exponentially in $|t|$, we conclude that $A_E \equiv A_G$ and hence that $U_E \equiv U_G$. This completes the proof of Theorem 5.4. ∎

COROLLARY 5.7. *The spectrum of* A *over* $\mathcal{H}_G$ *is the union of the spectrum of* A *over* $\mathcal{H}'_E$ *and a finite set consisting of the* $\pm\lambda_j$*'s and perhaps zero.*

PROOF. $\mathcal{H}'_G$ is a subspace of $\mathcal{H}_G$ which is invariant under A; it is complementary to $\mathcal{P}$, which is also invariant under A and which is finite dimensional. It is well known in operator theory that under these conditions the spectrum of A over $\mathcal{H}_G$ is the union of its spectrum over $\mathcal{H}'_G$ and its spectrum over $\mathcal{P}$.

According to Theorem 5.2, $\mathcal{J}$ is finite dimensional and by part (b) of Theorem 5.4 $\mathcal{J}$ is the null space of A. These null vectors of A on $\mathcal{H}'_G$ map into zero in $\mathcal{H}'_E$ and hence do not correspond to null vectors of A on $\mathcal{H}'_E$, which we again denote by A_E. On the other hand if A_E has a null vector, say, f in $\mathcal{H}'_E$ and if g in $\mathcal{H}'_G$ maps into f under the natural injection, then Ag but not g belongs to $\mathcal{J}$ so that $A^2_g = 0$ but $Ag \neq 0$; in this case $\lambda = 0$ is an eigenvalue of index two for A on $\mathcal{H}'_G$.

For values other than zero the spectra of A over $\mathcal{H}'_G$ and A_E coincide. For it is clear from part (c) of Theorem 5.4 that if λ belongs to the resolvent set of A over $\mathcal{H}'_G$ then it belongs to the resolvent set of A_E. Conversely if $\lambda \neq 0$ belongs to the resolvent set of A_E then we can solve

$$(\lambda - A)u = g \tag{5.31}$$

for any g in $\mathcal{H}'_G$. In fact if g maps into f in $\mathcal{H}'_E$ under the natural injection, then there exists a v in $\mathcal{H}'_E$ in the domain of A_E such that

$$(\lambda - A_E)v = f .$$

Choose any u′ in $\mathcal{H}'_G$ which maps into v; then

$$(\lambda - A)u' = g + j$$

for some j in $\mathcal{J}$. If we set

$$u = u' - \lambda^{-1} j ,$$

it is clear that u solves (5.31). Moreover $\lambda - A$ must be one-to-one because otherwise λ would be in the point spectrum of A_E by part (c) of Theorem 5.4. Since λ is in the resolvent set of A_E it follows that

$\lambda - A$ is one-to-one and onto $\mathcal{H}'_G$ and hence λ is in the resolvent set of A on $\mathcal{H}'_G$. A similar argument can be used to show for $\lambda \neq 0$ in the point spectrum of A_E that it is also in the point spectrum of A over $\mathcal{H}'_G$. This completes the proof of Corollary 5.7. ∎

We call the readers attention to the methods which we have used to solve the automorphic wave equation for data in $\mathcal{H}_G$; these methods were function theoretic depending on functional analysis and the theory of elliptic partial differential equations. Since the automorphic wave equation is hyperbolic, the theory of hyperbolic partial differential equations could also be employed to obtain an entirely different proof for the existence of $U(t)$ on $\mathcal{H}_G$.

§6. INCOMING AND OUTGOING SUBSPACES FOR THE AUTOMORPHIC WAVE EQUATION

In this section we define incoming and outgoing subspace $\mathfrak{D}_-$ and $\mathfrak{D}_+$ for the automorphic wave equation

$$u_{tt} = y^2 \Delta u + \frac{1}{4} u \,. \tag{6.1}$$

As we saw in Section 4, the operator on the right in (6.1) has a finite number of positive eigenvalues. Systems of this kind were analyzed in Section 3; here we shall show how the present system fits into that abstract framework. In particular we shall show that the incoming subspaces verify conditions (i) and (ii) of (2.1) as well as the modified condition (iii).

The incoming and outgoing waves which figure in this section are solutions of (6.1) which are independent of x. A solution u of (6.1) which is independent of x satisfies

$$u_{tt} = y^2 u_{yy} + \frac{1}{4} u \,. \tag{6.2}$$

The change of variables

$$s = \log y \,, \quad v = u/\sqrt{y} \tag{6.3}$$

changes (6.2) into the classical wave equation

$$v_{tt} = v_{ss} \,; \tag{6.4}$$

we leave the verification of this fact to the reader.

The energy associated with (6.4) is simply

$$\int (v_t^2 + v_s^2)\, ds \,. \tag{6.5}$$

We claim that under the transformation (6.3) this corresponds to the energy associated with (6.2):

$$E = \int \left(\frac{u_t^2}{y^2} + u_y^2 - \frac{1}{4y^2} u^2 \right) dy \ . \tag{6.6}$$

To verify this we substitute $u = \sqrt{y}\, v$ into (6.6):

$$E = \int \left(\frac{v_t^2}{y} + \left(\sqrt{y}\, v_y + \frac{1}{2} \frac{v}{\sqrt{y}} \right)^2 - \frac{v^2}{4y} \right) dy$$

$$= \int \left(\frac{v_t^2}{y} + y v_y^2 + v_y v \right) dy \ .$$

Assuming u to be zero at both ends of the integration we get

$$\int v_y v \, dy = \frac{v^2}{2} \Big| = 0 \ .$$

Substituting $s = \log y$ in the remaining integral we obtain (6.5).

The general solution of the wave equation (6.4) is of the form

$$v = \ell(s+t) + r(s-t) \ . \tag{6.7}$$

The first term on the right corresponds to a wave traveling to the left, the second term to a wave traveling to the right. An incoming solution is a wave traveling to the left which for times $t < 0$ is zero for $s < \log a$; similarly an outgoing solution is a wave traveling to the right which for $t > 0$ is zero for $s < \log a$. This is the same a that was introduced in Section 5 to define F_1.

Using the transformation (6.3) we can extend these notions to solutions of equation (6.2):

The general solution of (6.2) is of the form

$$\sqrt{y}\,\ell(\log y + t) + \sqrt{y}\,r(\log y - t) .$$

Let's make a logarithmic change of variables

$$\phi(y) = \ell(\log y) \tag{6.8$_-$}$$

or

$$\phi(y) = r(\log y) ; \tag{6.8$_+$}$$

we then get the following formulas for incoming and outgoing solutions:

$$u(w,t) = \sqrt{y}\,\phi(y e^{t}) , \tag{6.9$_-$}$$

$$u(w,t) = \sqrt{y}\,\phi(y e^{-t}) . \tag{6.9$_+$}$$

We take ϕ to be a C^∞ function which is zero for $y < a$. Then the function $u(x,t)$ defined by $(6.9)_-$ is zero for $y < a$ when $t \le 0$; similarly, u given by $(6.9)_+$ is zero for $y < a$ when $t \ge 0$. Thus the solution $(6.9)_-$ satisfies the boundary condition (4.5) for $t \le 0$, and $(6.9)_+$ satisfies it for $t \ge 0$, and so they are solutions of the automorphic wave equation for $t \le 0$, respectively ≥ 0. $(6.9)_-$ defined for $t \le 0$, is called an *incoming solution*, and $(6.9)_+$, defined for $t \ge 0$ is called *an outgoing* solution.

We define the incoming and outgoing subspaces $\mathfrak{D}_\pm$ *as the closure in* $\mathcal{H}_G$ *of the initial data of incoming and outgoing solutions respectively*:

$$(6.10) \qquad \begin{aligned} \mathfrak{D}_- &: \text{closure of } \{y^{1/2}\phi(y),\ \ y^{3/2}\phi'(y)\} , \\ \mathfrak{D}_+ &: \text{closure of } \{y^{1/2}\phi(y),\ -y^{3/2}\phi'(y)\} ; \end{aligned}$$

ϕ in C_0^∞, zero for $y < a$.

LEMMA 6.1. a) $\mathfrak{D}_\pm$ *satisfy properties* (i) *and* (ii) *of* (2.1)

b) $\mathfrak{D}_+$ *and* $\mathfrak{D}_-$ *are* E-*orthogonal.*

PROOF. Since incoming and outgoing solutions of (6.1) correspond to incoming and outgoing solutions of (6.4), we may as well verify properties (i) and (ii) for the latter, which is utterly obvious. Of course it is not much harder to verify them directly for the solutions $(6.9)_{\pm}$.

This takes care of part (a); for part (b) it is decidedly simpler to carry out the calculation for solution (6.7) of (6.4). Using the definition (6.5) of energy we get the following expression for the energy scalar product of a left traveling solution $\ell(s+t)$ and of a right traveling solution $r(s-t)$:

$$E(\ell, r) = \int (\ell_t r_t + \ell_s r_s)\, ds$$

$$= \int (-\ell' r' + \ell' r')\, ds = 0 \,.$$

This completes the proof of part (b). ■

By definition (6.9), data in $\mathcal{D}_{\pm}$ are zero for $y < a$. Hence for data in $\mathcal{D}_{\pm}$ the quadratic form K defined by (5.5) is zero. Since 2K was the amount we added to E to obtain the positive form G, it follows that for data d in $\mathcal{D}_+ + \mathcal{D}_-$ the energy form and the G-form are equal. By the same reasoning we get this more general result:

LEMMA 6.2.

$$E(d, g) = G(d, g)$$

for any d *in* $\mathcal{D}_- + \mathcal{D}_+$ *and any* g *in* $\mathcal{H}_G$.

REMARK 1. For the classical wave equation (6.4) on $\mathbf{R}_+$ and the associated incoming and outgoing spaces consisting of waves $\ell(s+t)$ and $r(s-t)$ going to the left and to the right, respectively, *the functions* ℓ *and* r *themselves furnish the translation representation*; more precisely, the representers are

$$\sqrt{2}\,\ell'(-s) \quad \text{and} \quad \sqrt{2}\,r'(s)\ .$$

The factor $\sqrt{2}$ is needed to make the representation unitary with respect to the energy norm. Using relations $(6.8)_\pm$ we can express these representers in terms of the function ϕ entering formulas (6.9) and (6.10); they are, respectively,

$$\sqrt{2}\, e^{-s} \phi'(e^{-s}) \tag{6.11$_-$}$$

and

$$\sqrt{2}\, e^{s} \phi'(e^{s})\ . \tag{6.11$_+$}$$

The principal task of this section will be to establish property (iii) of (2.1) for the incoming and outgoing subspaces $\mathfrak{D}'_\pm$ and $\mathfrak{D}''_\pm$ related to $\mathfrak{D}_\pm$ as in Section 3. We begin with a few definitions and lemmas.

We denote the zero Fourier coefficient of f with respect to x by $f^{(0)}$:

$$f^{(0)}(y) = \int_{-1/2}^{1/2} f(x,y)\,dx\ , \quad y > a\ . \tag{6.12}$$

LEMMA 6.3. *Let* $f = \{f_1, f_2\}$ *be an element of* $\mathcal{H}_G$ *which is E-orthogonal to* $\mathfrak{D}_-$; *then the following relation holds*:

$$f_2^{(0)} = -\,y^{3/2}\,\partial_y\left(\frac{f_1^{(0)}}{\sqrt{y}}\right). \tag{6.13$_-$}$$

Similarly, if f *is E-orthogonal to* $\mathfrak{D}_+$, *then*

$$f_2^{(0)} = y^{3/2}\,\partial_y\left(\frac{f_1^{(0)}}{\sqrt{y}}\right). \tag{6.13$_+$}$$

PROOF. Formula (6.10) gives a generic element of $\mathcal{D}_-$; using formula (5.3)′ for the E-norm, and the fact that data in $\mathcal{D}_-$ are zero for $y < a$ we can express E-orthogonality of f to $\mathcal{D}_-$ as follows:

$$\iint \left(y\, \partial_y \left(\frac{f_1}{\sqrt{y}} \right) \phi' + \frac{f_2 \phi'}{\sqrt{y}} \right) dx\, dy = 0$$

after performing the x integration the result can be expressed in terms of the notation (6.12) as

$$\iint \left(y\, \partial_y \left(\frac{f_1^{(0)}}{\sqrt{y}} \right) + \frac{f_2^{(0)}}{\sqrt{y}} \right) \phi' dy = 0 .$$

Since this is to hold for all ϕ with compact support, we conclude that the function multiplying ϕ' under the integral is constant:

$$y\, \partial_y \left(\frac{f_1^{(0)}}{\sqrt{y}} \right) + \frac{f_2^{(0)}}{\sqrt{y}} = \text{const}. \tag{6.14}$$

Divide by $\sqrt{y}$:

$$\sqrt{y}\, \partial_y \left(\frac{f_1^{(0)}}{\sqrt{y}} \right) + \frac{f_2^{(0)}}{y} = \frac{\text{const.}}{\sqrt{y}} . \tag{6.15}$$

For f in $\mathcal{H}_G$ both terms on the left are square integrable; therefore so is the right side; this is the case only if the constant is zero. Relation $(6.13)_-$ follows from this; we deduce $(6.13)_+$ analogously. ■

We say that f belongs locally to $\mathcal{H}_G$ if for every smooth function $\xi(y)$ with compact support in (a, ∞), ξf belongs to $\mathcal{H}_G$.

COROLLARY 6.4. *Let* $f = \{f_1, f_2\}$ *belong locally to* $\mathcal{H}_G$, *and suppose that* f *is* E-*orthogonal to all elements of* $\mathcal{D}_-$ *of form* $(6.10)_-$ *and of compact support. Then* f *satisfies*

$$(6.16)_- \qquad f_2^{(0)} = -y^{3/2}\partial_y\left(\frac{f_1^{(0)}}{\sqrt{y}}\right) + \text{const.}\ \sqrt{y}\ .$$

Similarly, if f *is* E-*orthogonal to all elements of* $\mathfrak{D}_+$ *of form* $(6.10)_+$ *and of compact support, then* f *satisfies*

$$(6.16)_+ \qquad f_2^{(0)} = y^{3/2}\partial_y\left(\frac{f_1^{(0)}}{\sqrt{y}}\right) + \text{const.}\ \sqrt{y}\ .$$

The proof goes the same way as that of Lemma 6.3, except that now we do not conclude that the constant in (6.15) is zero.

LEMMA 6.5. *Suppose that* f *is in* $\mathcal{H}_G$ *and that* f *is* E-*orthogonal to both* $\mathfrak{D}_-$ *and* $\mathfrak{D}_+$. *Then* f *is of the form*

$$(6.17) \qquad f^{(0)} = \{c\sqrt{y}, 0\} \quad \textit{for} \quad y > a\ .$$

PROOF. In this case f has to satisfy both $(6.13)_-$ and $(6.13)_+$. This is possible only if $f_2^{(0)} = 0$ and $f_1^{(0)}/\sqrt{y}$ is constant. This proves (6.17). ■

REMARK. If the data f is merely locally in $\mathcal{H}_G$ and E-orthogonal to all elements of both $\mathfrak{D}_-$ and $\mathfrak{D}_+$ of compact support, then f satisfies both $(6.16)_-$ and $(6.16)_+$. This is the case if and only if

$$f_1^{(0)} = c_1\sqrt{y}\log y + c_2\sqrt{y} \quad \text{and} \quad f_2^{(0)} = c_3\sqrt{y}\ .$$

The resolvent set of A in $\mathcal{H}_G$ is the set of λ for which

$$(\lambda - A) : u \to f$$

maps the domain of A one-to-one onto $\mathcal{H}_G$. Lemma 5.6 shows that the resolvent set of A is not empty, in fact that every real λ in absolute value > 4 belongs to it.

We denote by $\mathcal{D}^{\perp}$ the E-orthogonal complement of $\mathcal{D}_- + \mathcal{D}_+$ in $\mathcal{H}_G$.

THEOREM 6.6. *For every* λ *in the resolvent set of* A *in* $\mathcal{H}_G$, $(\lambda - A)^{-1}$ *maps the unit ball* $\{G(f) \le 1\}$ *in* $\mathcal{D}^{\perp}$ *into a compact subset of* $\mathcal{H}_G$.

REMARK. It suffices to prove the result for a single λ; in what follows it is convenient to take λ to be positive.

Theorem 6.6 is a consequence of a similar but simpler result:

LEMMA 6.7. *Let* ϕ *be any* C_0^∞ *function on* $\mathbf{R}_+$. *Define the operator* $M = M_\phi$ *by*

$$M_\phi = \int_0^\infty \phi(t)\, U(t)\, dt \ . \tag{6.18}$$

Then M_ϕ *maps the unit ball,* $G(f) \le 1$, *in* $\mathcal{D}^{\perp}$ *into a compact subset of* $\mathcal{H}_G$.

In order to deduce Theorem 6.6 from Lemma 6.7, we choose λ_0 so large that $\exp(-\lambda_0 t)\, \|U(t)\|$ remains bounded in t. Then we take $\lambda > \lambda_0$ and choose a sequence ϕ_n in $C_0^\infty(\mathbf{R}_+)$ so that $\phi_n(t)$ tends pointwise to $\exp(-\lambda t)$ and

$$\lim_{n \to \infty} \int_0^\infty |\phi_n(t) - e^{-\lambda t}|\, e^{\lambda_0 t}\, dt = 0 \ .$$

It then follows that the sequence of operators M_{ϕ_n} tends to $(\lambda - A)^{-1}$ in the sense of the operator norm in $\mathcal{H}_G$. Since the uniform limit of compact operators is compact, Theorem 6.6 follows from the compactness of the M_{ϕ_n} on $\mathcal{D}^{\perp}$.

We turn now to a proof of Lemma 6.7. We base it on the following two propositions:

a) The set

$$u = Mf\,, \qquad G(f) \le 1 \tag{6.19}$$

is compact with respect to the norm

$$G_Y(u) \equiv \iint_{F(Y)} (u_{1x}^2 + u_{1y}^2 + u_1^2 + u_2^2)\,dx\,dy \tag{6.20}$$

for any Y; here $F(Y)$ denotes the domain

$$F(Y) \equiv F \cap \{y < Y\}\ . \tag{6.21}$$

b) Given any ε, there is a $Y > a$ such that

$$G^Y(u) \equiv \iint_{Y<y} \left(u_{1x}^2 + y\left[\partial_y\left(\frac{u_1}{\sqrt{y}}\right)\right]^2 + \frac{u_2^2}{y^2}\right) dx\,dy \le \varepsilon\, G(f) \tag{6.22}$$

for all u of the form (6.19) with f in $\mathcal{D}^{\perp}$.

It follows from inequality (5.7)′ that for all u in $\mathcal{H}_G$

$$G(u) \le c(Y)\, G_Y(u) + G^Y(u)\ ,$$

where G_Y and G^Y are defined by (6.20) and (6.22), respectively, and c is a constant depending on Y. Using this fact and propositions a) and b) we can easily deduce Lemma 6.7.

The proof of part (a) goes as follows: We abbreviate Mf by u:

$$u = Mf = \int_0^\infty \phi(t)\,U(t)\,f\,dt\ . \tag{6.23}$$

Applying $(A^2 - I)$ to both sides of (6.23) and making use of the differential equation satisfied by U:

$$AU = \partial_t U, \quad A^2U = \partial_t^2 U ;$$

we get, after an integration by parts, the following:

$$(6.24) \qquad (A^2 - I)u = \int \phi(A^2 - I)Uf\, dt = \int \phi(\partial_t^2 - I)Uf\, dt$$

$$= \int [(\partial_t^2 - I)\phi]Uf\, dt .$$

Since ϕ is of class C_0^∞, the G-norm of the right side is bounded by the G-norm of f:

$$(6.25) \qquad G((A^2 - I)u) \leq \text{const. } G(f) .$$

Now according to the definition of A

$$(A^2 - I)u = \{(L - I)u_1, (L - I)u_2\} .$$

$L - I$ is an elliptic operator for which the estimate expressed in Theorem 4.4 holds. Using this and property (5.7) of the G-norm, we obtain the estimate:

$$\iint_{F(Y)} [u_1^2 + \sum(\partial u_1)^2 + \sum(\partial^2 u_1)^2 + u_2^2 + \sum(\partial u_2)^2]\, dx\, dy \leq \text{const. } G((A^2 - I)u) .$$

Estimating the right side by (6.25) we deduce that

$$(6.26) \qquad \iint_{F(Y)} [u_1^2 + \sum(\partial u_1)^2 + \sum(\partial^2 u_1)^2 + u_2^2 + \sum(\partial u_2)^2]\, dx\, dy \leq \text{const. } G(f) .$$

According to a classical compactness principle of Rellich already used in Section 4, a set of functions whose square integrals and that of their first derivatives over the compact, smoothly bounded domain F(Y) are uniformly bounded forms a compact set with respect to the L_2 norm over F(Y). According to inequality (6.26) we can apply this to the set of functions $\{u_1, \partial u_1, u_2\}$ where $u = \{u_1, u_2\} = M_\phi f$, $G(f) \leq 1$, and so we conclude that this set of functions is compact in the norm (6.20), as asserted in part (a).

We turn now to the proof of part (b) of Lemma 6.7. By assumption f is E-orthogonal to both $\mathfrak{D}_-$ and $\mathfrak{D}_+$; it follows then from (6.17) that its zero Fourier coefficient $f^{(0)}$ is of the form

$$f^{(0)} = \{c\sqrt{y}, 0\}, \quad \text{where} \quad c = f^{(0)}(a)/\sqrt{a} .$$

If we assume that f satisfies the additional condition

$$f^{(0)}(a) = 0 ,$$

then it follows that

$$f^{(0)}(y) = 0 \quad \text{for all} \quad y > a . \tag{6.27}$$

According to basic principles of functional analysis, the restriction of a set by the imposition of a finite number of linear bounded constraints cannot render it precompact unless it is itself precompact. Now by (4.26) and (5.7)′, $f^{(0)}(a)$ is a bounded functional in the G-norm. It therefore suffices to prove Lemma 6.7 under the additional condition (6.27) imposed on f. We shall also assume that f has compact support. Since these f are dense, this further restriction does not limit the generality of the result.

Next we denote by T the length of the support of ϕ, that is

$$\phi(t) = 0 \quad \text{for} \quad t > T . \tag{6.28}$$

U(t) is the solution operator for the hyperbolic equation

$$v_{tt} = Lv . \tag{6.29}$$

The non-Euclidean sound speed for this equation is one; hence a signal which originates at a point below $y = 2a$ will not get above $y = 2a\,e^t$ during time t. According to (6.23) and (6.28), u is a superposition of values of $U(t)f$, $0 < t \le T$; it follows that values of u at points where $y > 2a\,e^T$ do not depend on values of f below $y = 2a$. This suggests that we split f as

$$f = f^{(1)} + f^{(2)}$$

so that:

$$f^{(1)} = 0 \quad \text{for} \quad y < a \ ,$$

$$f^{(2)} = 0 \quad \text{for} \quad y > 2a$$

and so that

$$G(f^{(1)}) \le \text{const. } G(f) \ ,$$

the constant being independent of f. Since part (b) of the proof is about values of u for large y, $f^{(2)}$ plays no role in what follows. We therefore need only consider $f^{(1)}$ and to simplify the notation we denote it by f; that is we assume that f satisfies

$$(6.30) \qquad f = 0 \quad \text{for} \quad y < a \ .$$

Data f may be regarded as defined in the fundamental domain F, or as defined in the whole Poincaré plane Π and automorphic there. In the latter case we must regard $U(t)$ as the solution operator for the wave equation (6.29) in $\Pi \times \mathbf{R}$. In what follows we shall let $U(t)$ act on data defined in Π *which are not automorphic*. For such data we define the E-form and the G-form as before, except that now the domain of integration is all of Π, not just F. To indicate this distinction we shall denote the latter by

$$E_\Pi(g) \quad \text{and} \quad G_\Pi(g) \ .$$

We note that for data of compact support E_Π is positive; this follows from (5.3)′ if we take $F_1 = \Pi$.

Since by (6.30) f is zero outside the cusp neighborhood, the fact that f is automorphic merely means that f is periodic in x:

(6.31) $$f(x+1, y) = f(x, y) .$$

With this in mind we construct a smooth partition of unity on $\mathbf{R}$ which is invariant under unit translations:

$$\sum p(x+n) = 1, \qquad p \text{ has compact support} .$$

Then

$$f(x, y) = \sum f(x, y)\, p(x+n) = \sum g(x+n, y)$$

where

(6.32) $$g = fp .$$

Obviously the support of g is contained in the support of p and so is bounded in x. It is easy to show, using (6.27), that

(6.33) $$G_{II}(g) \leq \text{const. } G(f) .$$

Substituting the decomposition (6.32) of f into (6.23) we obtain a corresponding decomposition of u:

(6.34) $$u(x, y) = \sum h(x+n, y)$$

where

(6.35) $$h = Mg = \int \phi(t)\, U(t)\, g \; dt .$$

Let v be a solution of (6.29); its zero Fourier coefficient $v^{(0)}$ satisfies the equation

$$v_{tt}^{(0)} = \left(y^2 \partial_y^2 + \frac{1}{4}\right) v^{(0)} \qquad \text{for} \qquad y > 0 .$$

It follows from (6.27) and (6.30) that the initial data for $v^{(0)}$ is zero and hence that $v^{(0)}(y,t) = 0$ for all $y > 0$. As a consequence for u defined by (6.23)

$$u^{(0)} = 0 \quad \text{for all} \quad y > 0 . \tag{6.36}$$

Using the definition of the zero Fourier coefficient and the representation (6.34) we see from (6.36) that

$$0 = \int_{-1/2}^{1/2} u(x,y)\,dx = \int_{-1/2}^{1/2} \sum h(x+n,y)\,dx = \int_{-\infty}^{\infty} h(x,y)\,dx .$$

Subtracting this from (6.34) we get

$$u(x,y) = \sum_{-\infty}^{\infty} h(x+n,y) - \int_{-\infty}^{\infty} h(x,y)\,dx . \tag{6.37}$$

We turn now to the function h defined by (6.35). As observed after the relation (6.32), g has bounded support in x, say contained in the strip $|x| < 1$. It follows that $U(t)g$ is supported in the *domain of influence* of this strip, that is in the set of points in Π whose non-Euclidean distance from this strip is $\leq t$. A brief calculation shows that this point set is

$$|x| \leq 1 + y \sinh t . \tag{6.38}$$

Hence it follows from (6.28), (6.35) and (6.38) that the support of h is in

$$|x| \leq 1 + y \sinh T . \tag{6.38$'$}$$

The operator L commutes with dilation; that is if we define

$$h_c(x,y) = h(cx,cy) , \tag{6.39}$$

then

$$Lh_c = (Lh)_c . \tag{6.40}$$

It is also easy to verify that the E-form is invariant under dilation:

(6.41) $$E_{\Pi}(h_c) = E_{\Pi}(h) .$$

It follows from (6.40) that if v is a solution of equation (6.29), so is v_c defined as

$$v_c(t,x,y) = v(t,cx,cy) .$$

Consequently the operator M, defined in (6.18) also commutes with dilation; that is

(6.42) $$Mg_c = (Mg)_c = h_c .$$

Suppose as before that the support of g is contained in the strip $|x| \leq 1$. Then the support of g_c is contained in $|x| \leq 1/c$. Therefore it follows from (6.38)′ and (6.42) that the support of h_c is contained in

(6.43) $$|x| \leq \frac{1}{c} + y \sinh T .$$

Analogously to (6.24) and (6.25) we can estimate $(A^2 - I)h_c$ in terms of g_c:

$$E_{\Pi}((A^2 - I)h_c) \leq \text{const. } E_{\Pi}(g_c) .$$

Using (6.41), the relation between E and G and (6.33) we get the following string of inequalities:

$$E_{\Pi}(g_c) = E_{\Pi}(g) \leq G_{\Pi}(g) \leq \text{const. } G(f) .$$

Combining this with the above inequality we obtain

(6.44) $$E_{\Pi}((A^2 - I)h_c) \leq \text{const. } G(f) .$$

Since the functions h_c have compact support in Π, we can use the non-negative form of E:

(6.45) $$E_{\Pi}(k) = \iint_{\Pi} \left[k_{1x}^2 + y\left(\partial_y \frac{k_1}{\sqrt{y}}\right)^2 + \frac{k_2^2}{y^2} \right] dx\,dy .$$

We apply this to $k = (A^2 - I)h_c$; by definition of A

$$k = (A^2 - I)h_c = \{(L-I)h_{1_c}, (L-I)h_{2_c}\} .$$

Substituting this into (6.44) and (6.45), we get

$$\iint_{\Pi} \left\{ [(L-I)h_{1_{c}x}]^2 + y\left[\partial_y \frac{(L-I)h_{1_c}}{\sqrt{y}}\right]^2 + \frac{[(L-I)h_{2_c}]^2}{y^2} \right\} dx\, dy \le \text{const. } G(f) .$$

Using elliptic estimates analogously to the ones stated in Theorem 4.4, we can get from this the following estimates:

$$(6.46) \qquad \iint_C [h^2_{1_{c}xxx} + h^2_{1_{c}xyy} + h^2_{2_{c}xx} + h^2_{2_{c}yy}]\, dx\, dy \le \text{const. } G(f) ,$$

for any compact subset C of Π.

By the definition of h_c in (6.39)

$$h_c(x/c, y/c) = h(x,y) .$$

Substituting this in (6.37) we obtain

$$(6.47) \qquad u(x,y) = \sum_{-\infty}^{\infty} h_c\left(\frac{x+n}{c}, \frac{y}{c}\right) - c \int_{-\infty}^{\infty} h_c\left(v, \frac{y}{c}\right) dv ;$$

here $v = x/c$ was introduced as a new variable of integration.

Next we prove a general inequality:

LEMMA 6.8. *Let* m *be any function defined on* $\mathbf{R}$ *whose* N^{th} *derivative is integrable,* $N > 1$. *Let* b *denote any real number. Then*

$$(6.48) \qquad \left| b \sum_{-\infty}^{\infty} m(bn) - \int_{-\infty}^{\infty} m(v)\, dv \right| \le \text{const. } b^N \int_{-\infty}^{\infty} |\partial_v^N m(v)|\, dv .$$

PROOF. According to the Poisson summation formula

$$b \sum_{-\infty}^{\infty} m(bn) = \sum_{-\infty}^{\infty} \tilde{m}\left(\frac{2\pi}{b} j\right) ,$$

where $\tilde{m}$ is the Fourier transform of m:

$$\tilde{m}(\xi) = \int m(v) e^{i\xi v} dv .$$

The term

$$\tilde{m}(0) = \int m(v) dv$$

and therefore

$$b \sum_{-\infty}^{\infty} m(bn) - \int m(v) dv = \sum_{j \neq 0} \tilde{m}\left(\frac{2\pi}{b} j\right) . \tag{6.49}$$

Integrating by parts the integral defining $\tilde{m}$ yields the estimate

$$|\tilde{m}(\xi)| \leq \frac{1}{|\xi|^N} \int |\partial_v^N m(v)| \, dv .$$

Using this estimate on the right in (6.49) we get, for $N > 1$, the inequality (6.48). ■

We shall apply this lemma to

$$m(v) = h_{2_c}\left(\frac{x}{c} + v, \frac{y}{c}\right), \qquad b = 1/c ,$$

where h_{2_c} denotes the second component of h_c. In view of (6.47) we obtain from (6.48) the inequality

$$|u_2(x,y)| \leq \frac{\text{const.}}{c^{N-1}} \int |\partial_v^N h_{2_c}(v, y/c)| \, dv .$$

Choosing $N = 2$ and making use of the Schwarz inequality we deduce that

$$(6.50) \qquad |u_2(x,y)|^2 \le \frac{\text{const.}}{c^2}\, \ell(y/c) \int |\partial_v^2 h_{2_c}(v, y/c)|^2\, dv \ ,$$

where $\ell(w)$ denotes the length of the support of $h_2(v,w)$ as a function of v. According to (6.43)

$$(6.51) \qquad \ell(w) \le \frac{1}{c} + \text{const. } w \ .$$

We now take $c > 1$ and y in the interval

$$c \le y \le 2c \ .$$

We then get from (6.51) that

$$(6.51)' \qquad \ell(y/c) \le \ell(2) \le \text{const.}$$

We integrate (6.50) over $c \le y \le 2c$, $|x| \le 1/2$. This gives

$$\int_c^{2c} \int_{-1/2}^{1/2} [u_2(x,y)]^2 \frac{dx\,dy}{y^2} \le \frac{\text{const.}}{c^2} \int_c^{2c} \int |\partial_v^2 h_{2_c}(v, y/c)|^2 \frac{dv\,dy}{y^2} \ .$$

Introducing $w = y/c$ on the right as a new variable of integration, we obtain

$$\int_c^{2c} \int [u_2(x,y)]^2 \frac{dx\,dy}{y^2} \le \frac{\text{const.}}{c^3} \int_1^2 \int |\partial_v^2 h_{2_c}(v, w)| \frac{dv\,dw}{w^2} \ .$$

Finally using (6.46) to estimate the right side gives

$$\int_c^{2c} \int_{-1/2}^{1/2} [u_2(x,y)]^2 \frac{dx\,dy}{y^2} \le \frac{\text{const.}}{c^3}\, G(f) \ .$$

Setting $c = 2^n$ and summing over n, it follows immediately from this inequality that

$$\iint_{y>Y} [u_2(x,y)]^2 \frac{dx\,dy}{y^2} \le \frac{\text{const.}}{Y^3} G(f) .$$

In a similar manner we can get

$$\iint_{y>Y} \left\{ u_{1x}^2 + y\left[\partial_y\left(\frac{u_1}{\sqrt{y}}\right)\right]^2 \right\} dx\,dy \le \frac{\text{const.}}{Y^3} G(f) .$$

From these we obtain uniform estimates for the quantity $G^Y(u)$ defined in (6.22). This completes the proof of part (b) of Lemma 6.7 and with it the proof of Theorem 6.6. ∎

The G-norm has no intrinsic interest; its role is a purely technical one, that is to make possible the norm estimates which went into the proof of Theorem 6.6. Next we show that whatever is true in $\mathcal{H}_G$ is also true in the energy norm.

For the next argument it is helpful to have a notation for the elements of the quotient space $\mathcal{H}_G/\mathcal{J}$; we shall denote the image of an element f in $\mathcal{H}_G$ under the natural injection into $\mathcal{H}_G/\mathcal{J}$ by $\hat{f}$. We recall from Theorem 5.4 that $\mathcal{J}$ is the nullspace of A; therefore we can transfer the action of A to $\mathcal{H}_G/\mathcal{J}$:

If $u = (\lambda-A)^{-1}f$, then $\hat{u} = (\lambda-A)^{-1}\hat{f}$. $\mathcal{J}$ is E-orthogonal to everything; therefore the E-form can be defined on $\mathcal{H}_G/\mathcal{J}$ so that

$$E(\hat{f},\hat{g}) = E(f,g) .$$

We shall denote the image of $\mathcal{D}_-$ and $\mathcal{D}_+$ in $\mathcal{H}_G/\mathcal{J}$ by $\hat{\mathcal{D}}_-$ and $\hat{\mathcal{D}}_+$. According to Theorem 6.6 the set of u of the form

(6.52) $\quad u = (\lambda-A)^{-1}f,\ f \text{ in } \mathcal{H}'_G, G(f) \le 1,\ E(f, \mathcal{D}_- + \mathcal{D}_+) = 0$

is contained in a compact set in $\mathcal{H}'_G$. It follows that the image of this set in $\mathcal{H}'_G/\mathcal{G}$ also is contained in a compact subset of $\mathcal{H}'_G/\mathcal{G}$; and clearly the image of the set (6.52) in $\mathcal{H}'_G/\mathcal{G}$ is the set of $\hat{u}$ of the form

$$(6.53) \qquad \hat{u} = (\lambda - A)^{-1}\hat{f}, \ \hat{f} \text{ in } \mathcal{H}'_G/\mathcal{G}, \ G(\hat{f}) \leq 1, \ E(\hat{f}, \hat{\mathcal{D}}_- + \hat{\mathcal{D}}_+) = 0 .$$

We now relate $\mathcal{H}_G/\mathcal{G}$ to $\mathcal{H}$. Corollary 5.3 shows that under the natural injection $\mathcal{H}'_G/\mathcal{G}$ and $\mathcal{H}'_E$ are homeomorphic and hence can be identified in this way. Also E is preserved under the natural injection. On the other hand

$$\mathcal{H}_G = \mathcal{P} + \mathcal{H}'_G \quad \text{and} \quad \mathcal{H} = \mathcal{P} + \mathcal{H}'_E ;$$

and since $\mathcal{P} \cap \mathcal{G} = \{0\}$, the above identification can be extended in the obvious way from all of $\mathcal{H}_G/\mathcal{G}$ to all of $\mathcal{H}$. It is clear that the values of E are preserved under this extended identification. Theorem 5.4, part (c), together with (3.14) shows us that the action of U(t) in $\mathcal{H}_G/\mathcal{G}$ goes over into the action of U(t) on $\mathcal{H}$ under this correspondence.

By Lemma 6.2, E is positive on $\mathcal{D}_\pm$ so that $\mathcal{D}_\pm \cap \mathcal{G} = \{0\}$. Consequently $\mathcal{D}_-$ and $\mathcal{D}_+$ are in one-to-one correspondence with $\hat{\mathcal{D}}_-$ and $\hat{\mathcal{D}}_+$, respectively and it should cause no confusion if we denote by $\mathcal{D}_-$ and $\mathcal{D}_+$ the subspaces in $\mathcal{H}$ which correspond to $\hat{\mathcal{D}}_-$ and $\hat{\mathcal{D}}_+$ in $\mathcal{H}_G/\mathcal{G}$.

Under the above identification, the compactness of (6.53) goes over to

THEOREM 6.9. *For any* λ *in the resolvent set of* A *over* $\mathcal{H}'_E$, *the set of data* u *of the form*

$$(6.54) \qquad u = (\lambda - A)^{-1}f, \ f \text{ in } \mathcal{H}'_E, \ E(f) \leq 1, \ E(f, \mathcal{D}_- + \mathcal{D}_+) = 0 ,$$

lies in a compact subset of $\mathcal{H}'_E$.

Since our derivation started with Theorem 6.6, we have to initially take λ in the resolvent set of A over $\mathcal{H}'_G$. According to Corollary 5.7 this guarantees that λ is in the resolvent set of A over $\mathcal{H}'_E$. However once we know Theorem 6.8 to be true for one choice of λ, it follows for all λ in the resolvent set of A.

We now show that the subspaces $\mathfrak{D}_-$ and $\mathfrak{D}_+$ of $\mathcal{H}$ can play the role assigned to these symbols in Section 3. To show that these subspaces of $\mathcal{H}$ are E-weakly closed it suffices to prove that $\mathfrak{D}_-$ and $\mathfrak{D}_+$ considered as subspaces of $\mathcal{H}_G$ are E-weakly closed modulo $\mathcal{J}$. By construction $\mathfrak{D}_-$ and $\mathfrak{D}_+$ are closed subspaces of $\mathcal{H}_G$ and therefore G-weakly closed. Suppose for a sequence $\{d_n\} \subset \mathfrak{D}_-$ that there is an e in $\mathcal{H}_G$ such that

$$\lim_n E(d_n, f) = E(e, f)$$

for all f in $\mathcal{H}_G$. According to Lemma 6.2 we can replace the E form on the left in this relation by G and since $\mathfrak{D}_-$ is G-weakly closed there exists an d in $\mathfrak{D}_-$ such that for all f in $\mathcal{H}_G$

$$E(e, f) = \lim_n G(d_n, f) = G(d, f) = E(d, f) \ ;$$

here we have used Lemma 6.2 in the right equality. It follows that e equals d modulo $\mathcal{J}$ as required. Moreover, it is clear from Lemma 6.1 that $\mathfrak{D}_-$ and $\mathfrak{D}_+$ considered as subspaces of $\mathcal{H}$ satisfy properties (i) and (ii) of (2.1) and are E-orthogonal.

We recall from Section 3 the definition of $\mathfrak{D}''_-$ and $\mathfrak{D}''_+$ as

$$\mathfrak{D}''_- = \mathfrak{D}_- \cap \mathcal{H}'_E, \quad \mathfrak{D}''_+ = \mathfrak{D}_+ \cap \mathcal{H}'_E \ . \tag{6.55}$$

As we saw in Section 3, these are subspaces of $\mathfrak{D}_\pm$ which have finite codimension. It follows from this that if in (6.54) we replace the requirement $E(f, \mathfrak{D}_- + \mathfrak{D}_+) = 0$ by the less restrictive one:

$$E(f, \mathfrak{D}''_- + \mathfrak{D}''_+) = 0 \ ,$$

the resulting larger set is still contained in a compact subset of $\mathcal{H}'_E$. We state this mild generalization as

THEOREM 6.9″. *For any* λ *in the resolvent set of* A *over* $\mathcal{H}'_E$, *the set of* u *of the form*

$$u = (\lambda - A)^{-1} f, \quad f \text{ in } \mathcal{H}'_E, \quad E(f) \leq 1, \quad E(f, \mathcal{D}''_- + \mathcal{D}''_+) = 0 \tag{6.56}$$

lies in a compact subset of $\mathcal{H}'_E$.

We come now to the main result of this section. The operator A on $\mathcal{H}'_E$ is skew self-adjoint; it may contain purely imaginary point eigenvalues (in Section 8 we show that it contains infinitely many of them). Following the notation introduced in Section 3 we denote the space spanned by the eigenfunctions from the point spectrum of A by $\mathcal{H}'_p$, and its orthogonal complement in $\mathcal{H}'_E$ by $\mathcal{H}'_c$. Both are invariant subspaces for U(t). According to Lemma 2.11, $\mathcal{D}'_\pm$ and $\mathcal{D}''_\pm \subset \mathcal{H}'_c$.

THEOREM 6.10. $\mathcal{D}''_-$ *and* $\mathcal{D}''_+$ *are incoming and outgoing subspaces for* U(t) *restricted to* $\mathcal{H}'_c$; i.e. *they have the following properties*:

0) $\mathcal{D}''_-$ *and* $\mathcal{D}''_+$ *are closed*

i) $U(t)\mathcal{D}''_- \subset \mathcal{D}''_-$ *for* $t \leq 0$
$U(t)\mathcal{D}''_+ \subset \mathcal{D}''_+$ *for* $0 \leq t$

(6.57) ii) $\wedge U(t)\mathcal{D}''_- = \{0\}$
$\wedge U(t)\mathcal{D}''_+ = \{0\}$

iii) $\overline{\vee U(t)\mathcal{D}''_-} = \mathcal{H}'_c = \overline{\vee U(t)\mathcal{D}''_+}$.

iv) $\mathcal{D}''_-$ *and* $\mathcal{D}''_+$ *are orthogonal.*

PROOF. Properties (0), (i) and (ii) follow directly from Lemma 3.7. Since $\mathcal{D}_-$ and $\mathcal{D}_+$ are E-orthogonal the same is true for their subspaces $\mathcal{D}''_-$ and $\mathcal{D}''_+$. There remains property (iii); its proof will be divided into several steps.

LEMMA 6.11. *Let* Q''_- *denote the* E*-orthogonal projection of* $\mathcal{H}'_E$ *onto* $\mathfrak{D}''_-$. *Then*

(6.58) $$\lim_{t\to\infty} Q''_- U(t)f = 0$$

for every f *in* $\mathcal{H}'_E$.

PROOF. We claim that the union of

(6.59) $$\mathcal{H}'_E \ominus U(-T)\mathfrak{D}''_- \quad \text{for all } T .$$

is dense in $\mathcal{H}'_E$. For suppose not; then there is some nonzero k in $\mathcal{H}'_E$ which is E-orthogonal to all the subspaces (6.59); such a k belongs to all subspace $U(-T)\mathfrak{D}''_-$, contrary to property (ii).

Take now any f which belongs to one of the subspaces (6.59); for such an f
$$E(U(T)f, \mathfrak{D}''_-) = E(f, U(-T)\mathfrak{D}''_-) = 0 ,$$

i.e. $U(T)f$ is orthogonal to $\mathfrak{D}''_-$. Take $t > T$; then, since by property (i) $U(T-t)\mathfrak{D}''_- \subset \mathfrak{D}''_-$, we get

$$E(U(t)f, \mathfrak{D}''_-) = E(U(T)f, U(T-t)\mathfrak{D}''_-) = 0 ;$$

that is, for $t > T$, $U(t)f$ is orthogonal to $\mathfrak{D}''_-$. For such an f

$$Q''_- U(t)f = 0 \qquad \text{for} \qquad t > T ,$$

so that (6.58) holds. Since the set of such f is dense in $\mathcal{H}'_E$, (6.58) holds for all f in $\mathcal{H}'_E$. ■

In what follows we shall fix λ to be some point in the resolvent set of A, and introduce the abbreviation

(6.60) $$(\lambda - A)^{-1} = R .$$

We restate Theorem 6.9″ using this notation:

The set

$$\{Rg;\ g \text{ in } \mathcal{H}'_E,\ E(g) \le 1,\ g \perp \mathcal{D}''_- \oplus \mathcal{D}''_+\} \tag{6.61}$$

lies in a compact subset of $\mathcal{H}'_E$.

We are now ready to prove property (iii), i.e. that the $U(t)\mathcal{D}''_+$ are dense in $\mathcal{H}'_c$. For suppose not; then there would exist a nonzero f in $\mathcal{H}'_c$ which is orthogonal to all $U(t)\mathcal{D}''_+$. We may as well normalize f with E-norm 1. Since U preserves the E-scalar product, for such an f, $U(t)f$ is orthogonal to $\mathcal{D}''_+$ for all t.

Define $g(t)$ as

$$g(t) = (I - Q''_-)\,U(t)f\ . \tag{6.62}$$

Since Q''_- is the orthogonal projection into $\mathcal{D}''_-$ it follows that $g(t)$ is orthogonal to $\mathcal{D}''_-$. On the other hand, since by (iv) $\mathcal{D}''_-$ is orthogonal to $\mathcal{D}''_+$, we see that

$$g(t) = U(t)f - Q''_-U(t)f \tag{6.62$'$}$$

is the difference of two elements both orthogonal to $\mathcal{D}''_+$. So we conclude:

$$g(t) \text{ \textit{is orthogonal to} } \mathcal{D}''_- \oplus \mathcal{D}''_+\ . \tag{6.63}$$

Since $U(t)$ preserves the E-norm and $(I-Q''_-)$ diminishes it, we have

$$E(g(t)) \le E(f) = 1\ . \tag{6.64}$$

The relations (6.63) and (6.64) show that $g(t)$ satisfies the restrictions imposed on g in (6.61); therefore we conclude that the elements

$$\{Rg(t)\}$$

lie in a compact set. By (6.62)′

$$Rg(t) = RU(t)f - RQ''_-U(t)f\ . \tag{6.65}$$

The second term $RQ''_{-}U(t)f$ is a continuous function of t which according to (6.58) of Lemma 6.11 tends to zero as $t \to \infty$. The values of such a function for $0 \le t$ form a compact set; since the values of $Rg(t)$ were shown to form a compact set we conclude from (6.65) that also

$$\{RU(t)f;\ t \ge 0\}$$

lies in a compact set. Since R commutes with $U(t)$, $RU(t)f = U(t)Rf$. Introducing the abbreviation

$$Rf = k \tag{6.66}$$

we can write the above result as follows:

$$U(t)k,\ t \ge 0 \text{ \textit{lies in a compact set}}. \tag{6.67}$$

Next we need the following result:

THEOREM 6.12. *Let* $U(t)$ *be a strongly continuous group of unitary operators; the set*

$$\{U(t)k,\ t \ge 0\} \tag{6.68}$$

is contained in a compact subset if and only if k *is a linear combination of eigenvectors from the point spectrum of* U.

Before presenting the proof of this theorem we show how to use it to complete the proof of property (iii). It follows from (6.67) combined with Theorem 6.12 that k lies in the space $\mathcal{H}'_p$ spanned by the eigenvectors of A. Recalling that R abbreviates $(\lambda - A)^{-1}$ we get from (6.66) that

$$f = (A - \lambda)k\ ;$$

from this we conclude that f too lies in $\mathcal{H}'_p$. But this is a contradiction to f being in $\mathcal{H}'_c$, the orthogonal complement of $\mathcal{H}'_p$. This contradiction was caused by denying that $U(t)\mathcal{D}''_{+}$ is dense in $\mathcal{H}'_c$. $\mathcal{D}''_{-}$ can be treated similarly.

To complete the argument we have to prove Theorem 6.12. For this we need the following classical theorem of Wiener:

THEOREM 6.13 (Wiener). *Let* dm *be a complex measure on* $\mathbf{R}$ *whose total variation on* $\mathbf{R}$ *is* $< \infty$, *and which contains no point mass. Denote the Fourier transform of this measure by* $\tilde{m}$. *Then* $\tilde{m}(t)$ *tends to zero as* $t \to \infty$ *in the root mean sense:*

$$\lim \frac{1}{T} \int_0^T |\tilde{m}(t)|^2 \, dt \to 0 \,. \tag{6.69}$$

We omit the proof of this theorem; see e.g. Proposition 2.1 of Chapter V in [14].

We turn now to the proof of Theorem 6.12.

Suppose first that $U(t)$ has a discrete spectrum for k; then by the Stone theorem we can write

$$U(t)k = \sum e^{i\lambda_j t} k_j \tag{6.70}$$

where $\Sigma \|k_j\|^2 = \|k\|^2$. It is clear from the boundedness of the components that for any finite sum the set

$$\left\{ \sum_1^n e^{i\lambda_j t} k_j ; \ t \geq 0 \right\}$$

is compact. Since in (6.70) the partial sums converge uniformly for all t, it follows that (6.70) is likewise compact.

To prove the converse we split k as $k = k_p + k_c$, where k_p is a linear combination of eigenvectors and k_c is orthogonal to all eigenvectors. Then

$$U(t)k = U(t)k_p + U(t)k_c \,.$$

By the previous argument $U(t)k_p$ is compact; since we are assuming $U(t)k$ to be compact, it follows that

$$C = \{U(t)k_c, \ t \geq 0\}$$

is compact. That means that we can cover C by a finite number of balls of radius ε. Denote the centers of these balls by y_i, $i = 1, \cdots, n$. Then for each t there is an index i such that

$$\|U(t)k_c - y_i\| < \varepsilon .$$

We deduce from this that

$$\|k_c\| - \varepsilon \leq \|y_i\| \leq \|k_c\| + \varepsilon ,$$

and that

$$|(U(t)k_c - y_i, y_i)| \leq \varepsilon\|y_i\| \leq \varepsilon\|k_c\| + \varepsilon^2 .$$

From this we see that

$$\begin{aligned} |(U(t)k_c, y_i)| &\geq (y_i, y_i) - |(U(t)k_c - y_i, y_i)| \\ &\geq (\|k_c\| - \varepsilon)^2 - \varepsilon\|k_c\| - \varepsilon^2 = \|k_c\|^2 - 3\varepsilon\|k_c\| . \end{aligned}$$

Choosing $\varepsilon < \frac{1}{6}\|k_c\|$ we conclude that

$$|(U(t)k_c, y_i)| \geq \frac{1}{2}\|k_c\|^2 . \tag{6.71}$$

For each t, at least one of the inequalities (6.71) holds; therefore

$$\sum_{i=1}^{n} |(U(t)k_c, y_i)|^2 \geq \frac{1}{4}\|k_c\|^4 .$$

Integrating with respect to T and dividing by T we get

$$\sum_{i=1}^{n} \frac{1}{T}\int_0^T |U(t)k_c, y_i)|^2 \geq \frac{1}{4}\|k_c\|^4 . \tag{6.72}$$

Now by Stone's theorem on the spectral representation of unitary groups,

$$(U(t)k_c, y) = \int e^{i\lambda t} d(E_\lambda k_c, y) , \tag{6.73}$$

where E_λ gives the resolution of the identity for U. Since k_c is orthogonal to all eigenfunctions, the measures

$$dm_i = d(E_\lambda k_c, y_i)$$

contain no point masses. The Fourier transform of dm_i is given by (6.73):

$$\tilde{m}_i(t) = (U(t)k_c, y_i) .$$

In terms of this (6.72) can be written as

$$\sum_{i=1}^{n} \frac{1}{T} \int_0^T |\tilde{m}_i(t)|^2 \, dt \geq \frac{1}{4} \|k_c\|^4 .$$

According to Wiener's theorem, see (6.69), each term on the left tends to zero. But this implies that $k_c = 0$; so $k = k_p$ is a linear combination of eigenvectors of U, as asserted in Theorem 6.12. The proof of Theorem 6.10 is now complete. ∎

As in Section 3 we define $\mathcal{D}'_-$ and $\mathcal{D}'_+$ to be the E-orthogonal projection of $\mathcal{D}_-$ and $\mathcal{D}_+$ into $\mathcal{H}'_E$. Referring to Lemma 3.7 we have

COROLLARY 6.14. *$\mathcal{D}'_-$ and $\mathcal{D}'_+$ are incoming and outgoing subspaces for $U(t)$ restricted to $\mathcal{H}'_c$, inasmuch as they satisfy properties* (0), (i), (ii) *and* (iii) *of* (6.57).

We note that (iii) follows from the analogous property of $\mathcal{D}''_\pm$.

REMARK. In general property (iv) does *not* hold for $\mathcal{D}'_\pm$.

We are now ready to reap some benefits:

THEOREM 6.15. *The spectrum of the operator* A *over* $\mathcal{H}'_E$ *consists of a continuous spectrum of multiplicity one, and a point spectrum which accumulates only at* ∞.

PROOF. The decomposition

$$\mathcal{H}'_E = \mathcal{H}'_p \oplus \mathcal{H}'_c$$

splits up $\mathcal{H}'_E$ as the orthogonal sum of two invariant subspaces of U. Since $\mathcal{D}''_+$ and $\mathcal{D}''_-$ are both contained in $\mathcal{H}'_c$, $\mathcal{H}'_p$ is orthogonal to $\mathcal{D}''_- \oplus \mathcal{D}''_+$. It follows then from Theorem 6.9″ that $(\lambda - A)^{-1}$ is compact on $\mathcal{H}'_p$. This proves that the point spectrum of A accumulates only at ∞.

We turn now to $\mathcal{H}'_c$; according to Theorem 6.10 $\mathcal{D}''_-$ is an incoming subspace for U(t) restricted to $\mathcal{H}'_c$. As observed in Section 2 the spectrum of A over $\mathcal{H}'_c$ is continuous and of uniform multiplicity on R. To determine the multiplicity we shall rather deal with $\mathcal{D}'_-$, which according to Corollary 6.14 also is incoming in the sense of (2.1). For $\mathcal{D}'_-$ we can give an explicit translation representation as follows:

Every element d′ of $\mathcal{D}'_-$ is of the form

$$d' = Q'd, \quad d \text{ in } \mathcal{D}_-, \tag{6.74}$$

where Q′ is the E-orthogonal projection onto $\mathcal{H}'_E$. For the elements d of $\mathcal{D}_-$ we have already constructed a translation representation in (6.11). We have shown in Lemma 3.8 that for d and d′ connected by (6.74)

$$E(d') = E(d) .$$

Therefore the translation representation for $\mathcal{D}_-$ can be transferred to $\mathcal{D}'_-$ and it will retain its unitary character. Also, since U(t) commutes with Q′, it remains a translation representation of $\mathcal{D}'_-$. The functions which in (6.11) represent elements of $\mathcal{D}_-$ are scalar valued functions; this shows that the multiplicity of the continuous spectrum on $\mathbf{R}$ is one. This completes the proof of Theorem 6.15. ∎

In Section 3 we have established in Theorems 3.2 and 3.3 the connection between the spectrum of A and of L. Drawing on those results and Theorems 4.1 and 6.15 we deduce

COROLLARY 6.16. L *has continuous spectrum of multiplicity one on* $\mathbf{R}_-$; *its negative eigenvalues accumulate only at* $-\infty$. *In addition* L *has a finite number of positive eigenvalues.*

Finally we have

THEOREM 6.17. *Let* $Z''(t)$ *denote the semigroup introduced in* (2.25) *of Section* 2:

$$Z''(t) = P''_+ U(t) P''_- ,$$

acting on

$$\mathcal{K}'' = \mathcal{H}'_c \ominus \mathcal{D}''_- \ominus \mathcal{D}''_+ .$$

Then the resolvent of the generator B'' *of* Z'' *is meromorphic in the whole complex plane.*

PROOF. The resolvent of B'' can be obtained from the Laplace transform of Z''. We get for $\lambda > 0$

$$(\lambda - B'')^{-1} = P''_+ (\lambda - A)^{-1} P''_- .$$

We claim that $(\lambda - B'')^{-1}$ is compact; to show this we have to demonstrate that it maps the unit ball in $\mathcal{K}''$ into a compact subset of $\mathcal{K}''$. Since $\mathcal{K}''$ is orthogonal to both $\mathcal{D}''_-$ and $\mathcal{D}''_+$, it follows from Theorem 6.9″ that $(\lambda - A)^{-1}$ maps the unit ball of $\mathcal{K}''$ into a compact set; the projection P''_+ carries this compact set into another compact set. The conclusion of the theorem follows now from the compactness of $(\lambda - B'')^{-1}$.

§7. THE SCATTERING MATRIX FOR THE AUTOMORPHIC WAVE EQUATION

This section is the culmination of the previous six; in it we derive a formula for the concrete realization of the generalized eigenfunctions $e(z)$ for the automorphic wave equation. From this concrete realization we obtain explicit formulas for the spectral representation and the scattering matrix.

For notational convenience the space $\mathcal{H}'_c$ will be abbreviated in this section simply as $\mathcal{H}$.

THEOREM 7.1. *The scattering matrix* $\mathcal{S}'(z)$ *for the automorphic wave equation is a meromorphic function.*

REMARK. We have shown in Theorem 6.15 that for the modular group the continuous spectrum of A has multiplicity one and therefore the scattering "matrix" is in this case a scalar function.

PROOF. By formula (3.43) the scattering matrix can be written in factored form as

$$\mathcal{S}' = \mathcal{S}_+^{-1}\mathcal{S}''\mathcal{S}_-^{-1} . \tag{7.1}$$

According to Theorem 6.17, the semigroup Z'' associated with the pair of orthogonal incoming and outgoing subspaces $\mathcal{D}''_-$ and $\mathcal{D}''_+$ is generated by an operator B'' which has a compact resolvent. As a consequence the resolvent of B'' is a meromorphic function. It follows then from Theorem 2.9 that the associated scattering matrix $\mathcal{S}''(z)$ too is a meromorphic function.

According to Theorem 3.9, the scattering matrices $\mathcal{S}_-$ and $\mathcal{S}_+$ are Blaschke products with poles in the upper half plane; it follows that $\mathcal{S}_-^{-1}$ and $\mathcal{S}_+^{-1}$ are finite Blaschke products with poles in the lower half plane:

$$\mathcal{S}_-^{-1}(z) = \mathcal{S}_+^{-1}(z) = \prod \left(\frac{z - i\lambda_j}{z + i\lambda_j}\right) \equiv \mathcal{B}(z) ,$$

where the λ_j are the relevant eigenvalues defined in Section 3. The conclusion of Theorem 7.1 therefore follows from formula (7.1). ∎

In Appendix 2 to Section 2 we have introduced the space of functions ℓ with finite λ-norm:

$$\|\ell\|_\lambda^2 = \int_{-\infty}^{0} |\ell(s)|^2 e^{2\lambda s} \, ds + \int_0^{\infty} |\ell(s)|^2 \, ds < \infty .$$

Given a translation representation of $\mathcal{H}$:

$$\mathcal{H} \leftrightarrow L^2(\mathbf{R}, \mathcal{N})$$

we define the λ-norm of an element f of $\mathcal{H}$ as the λ-norm of the function ℓ representing it, and we define $\mathcal{H}^\lambda$ as the completion of $\mathcal{H}$ in the λ-norm. In what follows, *we shall denote by* $\mathcal{H}^\lambda$ *the completion of* $\mathcal{H}$ *in the* λ*-norm formed with respect to the* $\mathcal{D}'_-$ *representation.* We take λ to be any positive number less than all of the relevant eigenvalues λ_j:

(7.2) $$0 < \lambda < \lambda_j .$$

Given any f in $\mathcal{H}$, we denote its $\mathcal{D}'_-$, $\mathcal{D}''_-$, $\mathcal{D}''_+$ and $\mathcal{D}'_+$ representers by ℓ'_-, ℓ''_-, ℓ''_+ and ℓ'_+. We claim that if λ satisfies (7.2), then

(7.3) $$\|\ell'_-\|_\lambda \geq k\|\ell''_-\|_\lambda \geq k\|\ell''_+\|_\lambda \geq k^2\|\ell'_+\|_\lambda .$$

PROOF. According to (7.1), the $\mathcal{D}'_-$ representation is linked to the $\mathcal{D}''_-$ representation by a scattering matrix $\mathcal{S}_-^{-1}$ that is the reciprocal of a

Blaschke product whose zeros are at $-i\lambda_j$. The first inequality in (7.3) follows then from Theorem 2.18. The second inequality in (7.3) follows from Theorem 2.13 since the $\mathcal{D}''_-$ representation is linked to the $\mathcal{D}''_+$ representation by a causal scattering operator. The third inequality is analogous to the first. ■

$\mathcal{H}^\lambda$ was defined as the completion of $\mathcal{H}$ in the λ-norm of the $\mathcal{D}'_-$ representation. It follows by continuity from (7.3) that

THEOREM 7.2. *We can assign to every element of* $\mathcal{H}^\lambda$ *a* $\mathcal{D}''_-$, $\mathcal{D}''_+$ *and* $\mathcal{D}'_+$ *representation with finite* λ*-norm; the dependence of these representations on* f *is continuous in the* $\mathcal{H}^\lambda$ *norm.*

Before proceeding further we have to review briefly a few needed formulas from Sections 3 and 6. Using these we shall construct concrete realizations of elements of $\mathcal{H}^\lambda$ as data on F, with locally finite but infinite total energy.

Recall that $\mathcal{D}'_-$ was defined in (3.29) as the projection of $\mathcal{D}_-$ into $\mathcal{H}'$. It was shown in Lemma 3.5, see formula (3.28), that elements d'_- of $\mathcal{D}'_-$ are of the form

$$d'_- = d_- + \sum c_j^- f_j^+ , \tag{7.4}$$

where d_- is any element of $\mathcal{D}_-$, f_j^+ is defined by formula (3.12) and the coefficients c_j^- are chosen so that d'_- is orthogonal to all the f_k^-, also defined in (3.12). In view of relations (3.17) we have the following formula for c_j^-:

$$c_j^- = E(d_-, f_j^-)/\lambda_j^2 . \tag{7.5$_-$}$$

For the concrete construction of the translation representations in the present situation we have to use the results of Section 6. According to formula (6.10), $\mathcal{D}_-$ and $\mathcal{D}_+$ consist of data of the following respective form

$$(7.6)_- \qquad d_- = \{y^{1/2}\phi_-(y), y^{3/2}\psi_-(y)\}$$

and

$$(7.6)_+ \qquad d_+ = \{y^{1/2}\phi_+(y), -y^{3/2}\psi_+(y)\} .$$

Here ψ is an abbreviation for

$$(7.7) \qquad \psi(y) = \partial_y \phi(y) ,$$

and the functions $\phi_\pm(y) = 0$ for $y < a$. It follows then from (7.4) that elements of $\mathfrak{D}'_-$ and $\mathfrak{D}'_+$ are of the form

$$(7.6)'_- \qquad d'_- = \{y^{1/2}\phi_-(y), y^{3/2}\psi_-(y)\} + \sum c_j^- f_j^+$$

$$(7.6)'_+ \qquad d'_+ = \{y^{1/2}\phi_+(y), -y^{3/2}\psi_+(y)\} + \sum c_j^+ f_j^- ,$$

where the constants $c_j^\pm$ are determined by formula $(7.5)_-$ and its plus analogue $(7.5)_+$.

As we have already noted (see the discussion following formula (6.74)), the translation representer for $d'_\pm$ can be taken over from that of $d_\pm$. Making use of (6.11) we see that the translation representer of an element of form $(7.6)'_\pm$ is

$$(7.8)_\pm \qquad d'_\pm(s) = \sqrt{2}\, a e^{\pm s} \psi_\pm(a e^{\pm s}) .$$

These relations can be inverted to express ψ in terms of $d'_\pm$:

$$(7.9)_- \qquad \psi_-(y) = \frac{1}{\sqrt{2}\, y} d'_-(\alpha - \log y)$$

$$(7.9)_+ \qquad \psi_+(y) = \frac{1}{\sqrt{2}\, y} d'_+(\log y - \alpha) ;$$

here $\alpha = \log a$.

We summarize these formulas as a

PROPOSITION. *The function* $d'_-(s)$ *in* L_2 *whose support lies on* $\mathbf{R}_-$ *is the* $\mathfrak{D}'_-$ *translation representer of the element* d'_- *given by formula* (7.6)$'_-$, *where* ψ_- *is defined by* (7.9)$_-$, ϕ_- *is defined by* (7.7) *and* c_j^- *is defined by* (7.5)$_-$.

We now extend this proposition to any function d_- *supported on* $\mathbf{R}_-$ *which has finite* λ*-norm.* We claim that this definition makes sense, i.e. that the integral in (7.5)$_-$ defining c_j^- converges. To see this we note that $\|d'_-\|_\lambda < \infty$ and definition (7.9)$_-$ imply that

$$\int |\psi_-(y)|^2 y^{1-2\lambda}\,dy < \infty . \tag{7.10}$$

Moreover the zero Fourier coefficient of $f_j^\pm$ has the following form:

$$f_j^{\pm(0)} = \text{const.}\,\{y^{1/2-\lambda_j}, \pm\lambda_j y^{1/2-\lambda_j}\} . \tag{7.11}$$

To deduce this we recall from formulas (3.11) and (3.12) of Section 3 that $f_j^\pm$ is of the form

$$f_j^\pm = \{q_j, \pm\lambda_j q_j\} ,$$

where q_j is an eigenfunction of L:

$$Lq_j = \lambda_j^2 q_j .$$

Integrating this with respect to x gives the following differential equation for $q_j^{(0)}$:

$$y^2 \partial_y^2 q_j^{(0)} + \frac{1}{4} q_j^{(0)} = \lambda_j^2 q_j^{(0)} .$$

This equation has two solutions, $y^{1/2-\lambda_j}$ and $y^{1/2+\lambda_j}$; only the first is in L_2, which proves (7.11).

Since d_- as defined by (7.6)$'_-$ is independent of x, we can rewrite formula (7.5)$_-$ for c_j^- as follows:

$$c_j^- = E(d_-, f_j^-) = E(d_-, f_j^{-(0)}) .$$

Using (7.10) and (7.11), plus the fact that $\lambda < \lambda_j$ we conclude that the above integral converges absolutely.

REMARK. The coefficients c_j^- are bounded linear functionals of d'_- in the λ-norm.

Given any ℓ with finite λ-norm, we can break it up as

$$\ell = \ell_1 + \ell_2 \tag{7.12}$$

where $\ell_1(s)$ is supported in $\mathbf{R}_-$ and $\ell_2(s)$ vanishes for $s < -1$. It follows that ℓ_1 has finite λ-norm and that ℓ_2 is in L_2; therefore each has a concrete realization. Denoting these concrete realizations by f_1 and f_2, *we now define*

$$f = f_1 + f_2 \tag{7.12$'$}$$

as the concrete realization of ℓ.

REMARK. The breakup (7.12) is not unique; it is easy to show, but we leave it to the reader, that the above definition of f does not depend on the particular decomposition employed in (7.12).

The extended space $\mathcal{H}^\lambda$ shares most of the properties of the space $\mathcal{H}$. A useful result is the following:

LEMMA 7.3. *Suppose* f *in* $\mathcal{H}^\lambda$ *is represented by* $\ell(s)$ *and both* ℓ *and* $\partial_s \ell$ *have finite* λ*-norm; then* f *belongs to the domain of the differential operator* A *and* Af *is represented by* $-\partial_s \ell$.

PROOF. It suffices to prove this for those ℓ with support in $\mathbf{R}_-$. In this case it follows from formula $(7.9)_-$ that ψ_- is once, and so ϕ_- is twice differentiable. This shows that f lies in the domain of A; an explicit calculation of Af from (7.6) gives

$$\mathrm{Af} = \{y^{3/2}\psi_-, y^{5/2}\partial_y^2\phi_- + y^{3/2}\partial_y\phi_-\} + \sum c_j^- \lambda_j f_j^+$$

$$= \{y^{1/2}(y\psi_-), y^{3/2}\partial_y(y\psi_-)\} + \sum c_y^- \lambda_j f_j^+ .$$

Multiplying (7.9)_ by y and applying ∂_y we conclude that

$$\partial_y(y\psi_-) = \frac{1}{\sqrt{2}\,y}(\partial_s d'_-)(a - \log y) .$$

Using (7.9)_ once more with d'_- replaced by $\partial_s d'$ we conclude, on the basis of the above expression for Af, that Af is represented by $\partial_s d'_-$. This completes the proof of Lemma 7.3. ■

Next we shall derive a formula for the $\mathfrak{D}'_+$ component of an element of $\mathcal{H}$ and show that this formula makes sense also for elements of $\mathcal{H}^\lambda$.

Given f in $\mathcal{H}$, we write its zero Fourier component in the following form:

(7.13) $$f^{(0)} = \{y^{1/2}\alpha, y^{3/2}\beta\} ,$$

α and β functions of y.

ASSERTION. $f^{(0)}$ *can be decomposed as*

(7.13)′ $$f^{(0)} = f_+ + f_- + k\{y^{1/2}, 0\} ,$$

where f_+ *and* f_- *belong to* $\mathfrak{D}_+$ *and* $\mathfrak{D}_-$, *respectively, and* k *is a constant.*

PROOF. Elements of $\mathfrak{D}_-$ and $\mathfrak{D}_+$ are described by formulas (7.6) and (7.7); from them we get

$$f_- + f_+ = \{y^{1/2}(\phi_- + \phi_+), y^{3/2}(\psi_- - \psi_+)\} .$$

To obtain the decomposition (7.13)′ we must have

$$\alpha = \phi_- + \phi_+ + k$$

$$\beta = \psi_- - \psi_+ .$$

Differentiating the first relation with respect to y and using (7.7) we get

$$\partial_y \alpha = \psi_- + \psi_+ .$$

Combining this with the second relation we get

$$\psi_+ = \frac{1}{2}(\partial_y \alpha - \beta), \quad \psi_- = (\partial_y \alpha + \beta) . \tag{7.14}$$

Knowing $\psi_\pm$ we can, using (7.7) and the condition $\phi_\pm(a) = 0$, determine $\phi_\pm$; this fixes the constant k. ∎

LEMMA 7.4. *In terms of the decomposition* (7.13)′ *for* f *in* $\mathcal{H}$, *the* $\mathcal{D}'_-$ *and* $\mathcal{D}'_+$ *components of* f *are*:

(7.15)₋ $$f'_- = f_- + \sum a_j^- f_j^+$$

(7.15)₊ $$f'_+ = f_+ + \sum a_j^+ f_j^- ,$$

where the constants $a_j^\pm$ *are given by* $(7.5)_\pm$.

PROOF. Clearly the elements given by (7.15) belong to $\mathcal{D}'_\pm$. To show that they are the E-orthogonal projections of f we have to verify that $f - f'_-$ is E-orthogonal to $\mathcal{D}'_-$ and that $f - f'_+$ is E-orthogonal to $\mathcal{D}'_+$.

Let h'_- be any element of $\mathcal{D}'_-$; by $(7.6)'_-$, h'_- is of the form

$$h'_- = h_- + \sum b_j^- f_j^+, \quad h_- \text{ in } \mathcal{D}_- .$$

Since $f_j^\pm$ are orthogonal to $\mathcal{H}$, and since elements of $\mathcal{D}_-$ are independent of x, the following identities hold:

$$(7.16) \qquad E(f-f'_-,h'_-) = E(f-f'_-,h_-) = E(f^{(0)}-f'^{(0)}_-,h_-) .$$

It follows from (7.13)′ and $(7.15)_-$ that

$$f^{(0)}-f'^{(0)}_- = f_+ - \sum a_j^- f_j^{+(0)} + k\{y^{1/2},0\} .$$

According to (7.11) $f_j^{+(0)}$ is of the form

$$\{y^{1/2}y^{-\lambda_j}, -y^{3/2}\partial_y y^{-\lambda_j}\} .$$

Hence each of the terms in the above decomposition of $f-f'_-$ satisfies the relation $(6.13)_-$:

$$g_2^{(0)} = -y^{3/2}\partial_y\left(\frac{g_1^{(0)}}{\sqrt{y}}\right)$$

and is therefore by Lemma 6.3 E-orthogonal to $\mathfrak{D}_-$; we conclude by (7.16) that $f-f'_-$ is orthogonal to $\mathfrak{D}'_-$. This completes the proof that $(7.15)_-$ is the projection of f onto $\mathfrak{D}'_-$; that $(7.15)_+$ is the projection of f into $\mathfrak{D}'_+$ can be verified similarly. ■

LEMMA 7.5. *The projections of* f *into* $\mathfrak{D}'_+$ *given by formula* $(7.15)_+$ *can be extended to any* f *in* $\mathcal{H}^\lambda$, *and the dependence of* f'_+ *on* f, *regarded as mapping from* $\mathcal{H}^\lambda$ *to* $\mathcal{H}$, *is continuous.*

PROOF. As remarked already, every f in $\mathcal{H}^\lambda$ can be decomposed as in (7.12)′, i.e. as the sum of an element f_2 in $\mathcal{H}$, whose $\mathfrak{D}'_+$ projection is obviously continuous, plus another f_1 whose $\mathfrak{D}'_-$ translation representer is supported on $\mathbf{R}_-$. The latter element was defined by formula $(7.6)'_-$; the zero Fourier component of such an element is of the form

$$\{y^{1/2}\alpha, y^{3/2}\beta\}$$

where, using (7.7) and (7.11) we have

$$\alpha = \phi_- + \sum c_j^- y^{-\lambda_j}$$

$$\beta = \partial_y \phi_- + \sum \lambda_j c_j^- y^{-\lambda_j - 1} .$$

We wish to determine the $\mathfrak{D}'_+$ component of f_1. Substituting these values of α and β into (7.14) we get that

$$\psi_+ = -\sum \lambda_j c_j^- y^{-\lambda_j - 1} .$$

By integration we obtain

$$\phi_+ = \sum c_j^- (y^{-\lambda_j} - a^{-\lambda_j}) .$$

Having determined $d_+ = \{y^{1/2}\phi_+, -y^{3/2}\psi_+\}$ in this way, we then compute the projection of f_1 into $\mathfrak{D}'_+$ by means of $(7.15)_+$; the constants a_j^+ which appear in $(7.15)_+$ are determined by integration as in $(7.5)_+$. Finally the projection f'_+ of f into $\mathfrak{D}'_+$ is simply the sum of the projections of f_1 and f_2; it is an element of $\mathcal{H}$. That f'_+ as element of $\mathcal{H}$ depends continuously on f measured in the λ-norm follows from the fact, noted earlier, that the coefficients c_j^- and a_j^+ are bounded linear functionals of f. This completes the proof of Lemma 7.5. ∎

At the beginning of this section in Theorem 7.2 we showed how to assign to any f in $\mathcal{H}^\lambda$ a $\mathfrak{D}'_+$ representer ℓ'_+ which satisfies (7.3):

$$\|\ell'_+\|_\lambda \leq k^2 \|f\|_\lambda .$$

In term of ℓ'_+ we can define directly the $\mathfrak{D}'_+$ component of f as the element represented in the $\mathfrak{D}'_+$ representation by the function

$$(7.17) \qquad \begin{cases} 0 & \text{for} \quad s < 0 \\ \ell'_+(s) & \text{for} \quad s > 0 . \end{cases}$$

Since the L_2 norm of ℓ'_+ over $\mathbf{R}_+$ is $\leq \|\ell'_+\|_\lambda$, it follows from this and (7.3) that the $\mathcal{H}$-norm of the $\mathcal{D}'_+$ component of f as defined above is $\leq k^2 \|f\|_\lambda$. This shows that this way of defining the $\mathcal{D}'_+$ component of f is a continuous mapping of $\mathcal{H}^\lambda$ into $\mathcal{H}$. In Lemma 7.5 we have shown that defining the $\mathcal{D}'_+$ component by formula $(7.15)_+$ is also continuous. Since the two definitions agree when f belongs to $\mathcal{H}$, we conclude

LEMMA 7.6. *The two ways of defining the* $\mathcal{D}'_+$ *component of* f *in* $\mathcal{H}^\lambda$ *are the same.*

We turn now to the generalized eigenfunctions of U(t) in $\mathcal{H}^\lambda$. For any complex z satisfying

$$-\lambda < \operatorname{Im} z < 0 , \tag{7.18}$$

the function

$$e^{-izs}$$

has finite λ-norm; therefore it is the $\mathcal{D}'_-$ representative of an element e(z) in $\mathcal{H}^\lambda$. Since all elements of $\mathcal{H}^\lambda$ have concrete realizations as data defined in the fundamental domain F, so does e(z); we denote the concrete realization of e(z) by the same symbol.

According to Theorem 7.2 every element in $\mathcal{H}^\lambda$ has a $\mathcal{D}'_-$, $\mathcal{D}''_-$, $\mathcal{D}''_+$ and $\mathcal{D}'_+$ representer. For the elements e(z) these are particularly simple.

THEOREM 7.7. *The* $\mathcal{D}'_-$, $\mathcal{D}''_-$, $\mathcal{D}''_+$, $\mathcal{D}'_+$ *translation representers of* e(z) *are:*

$$e^{-izs}, \quad \mathcal{C}(z)e^{-izs}, \quad \mathcal{S}''(z)\mathcal{C}(z)e^{-izs}, \quad \mathcal{S}'(z)e^{-izs} ; \tag{7.19}$$

here $\mathcal{C}(z)$ *is the reciprocal of the Blaschke product* $\mathcal{S}_-(z)$.

PROOF. To pass from the $\mathcal{D}'_-$ translation representation to the $\mathcal{D}''_-$ representation, or from the $\mathcal{D}''_+$ representation to the $\mathcal{D}'_+$ representation, we have to perform a convolution with the function $C_-(s)$ or $C_+(s)$ obtained as in (2.93) from $\mathcal{B} = \mathcal{S}_-$ or $\mathcal{S}_+$, respectively. The convolution of C with e^{-izs} is simply $\mathcal{C}(z)e^{-izs}$. On the other hand the passage from the $\mathfrak{D}''_-$ representation to the $\mathfrak{D}''_+$ is accomplished by means of a causal scattering operator so that (2.14) applies. Putting these facts together as in the proof of Theorem 7.2 we deduce assertion (7.19). ∎

The projection operators P'_-, P'_+, P''_-, P''_+ are defined in $\mathcal{H}$ as the operators which remove the $\mathfrak{D}'_-$, $\mathfrak{D}'_+$, $\mathfrak{D}''_-$, $\mathfrak{D}''_+$ components respectively. These operators can be extended to $\mathcal{H}^\lambda$ for λ in the range (7.2); we note that P'_- and P''_- are continuous maps of $\mathcal{H}^\lambda$ into $\mathcal{H}$, whereas P'_+ and P''_+ are continuous only as maps of $\mathcal{H}^\lambda$ into $\mathcal{H}^\lambda$.

LEMMA 7.8.

$$P''_- e(z) + P''_+ e(z) - e(z)$$

can be continued as a meromorphic function whose values lie in $\mathcal{H}$.

PROOF. We have already observed in the course of proving Theorem 7.1 that according to Theorem 6.15 the semigroup Z'' associated with the pair of subspaces $\mathfrak{D}''_-$ and $\mathfrak{D}''_+$ is generated by the operator B'' which has a compact resolvent. Lemma 7.8 follows then from Theorem 2.10. ∎

LEMMA 7.9.

$$P'_- e(z) - P''_- e(z) \quad \textit{and} \quad P'_+ e(z) - P''_+ e(z)$$

can be continued as meromorphic functions whose values lie in $\mathcal{H}$.

PROOF. Since the $\mathfrak{D}''_-$ and $\mathfrak{D}'_-$ representations are related by a Blaschke product we can in Theorem 2.19 take T_1 and T_2 to be the $\mathfrak{D}''_-$ and $\mathfrak{D}'_-$ translation representations; the conclusion of the first part of Lemma 7.9 then follows from Theorem 2.19; the second part follows analogously. ∎

Combining Lemmas 7.8 and 7.9 we conclude

LEMMA 7.10.

$$P'_- e(z) + P'_+ e(z) - e(z)$$

can be continued from the strip (7.18) *into the whole plane as a meromorphic function whose values lie in* $\mathcal{H}$.

REMARK. The poles are located in $-i\sigma(B'')$ and the relevant $-i\lambda_j$.

To study $e(z)$ further we simply follow the steps of the general construction of elements of $\mathcal{H}^\lambda$:

We decompose $\ell(s) = e^{-izs}$ as in (7.12):

$$e^{-izs} = \ell_1(s) + \ell_2(s) ,$$

where

$$\ell_1(s) = e^{-izs} \quad \text{for} \quad s < -1 . \tag{7.20}$$

We define $\psi_-(y)$ by formula $(7.9)_-$:

$$\psi_-(y) = \frac{1}{\sqrt{2}\, y} \ell_1(a - \log y) .$$

Using (7.20) we get that

$$\psi_-(y) = \frac{a^{-iz}}{\sqrt{2}} y^{iz-1} \quad \text{for} \quad y > ae . \tag{7.21}$$

Guided by (7.21) we set for all y

$$\phi_-(y) = \frac{a^{-iz}}{\sqrt{2}\, iz} \xi(y) y^{iz} , \tag{7.22}$$

where $\xi(y)$ is a C^∞ cutoff function satisfying

(7.23)
$$\xi(y) = \begin{cases} 1 & \text{for} \quad y > ae \\ 0 & \text{for} \quad y < a \ . \end{cases}$$

Then we set, following (7.7), $\psi_-(y) = \partial_y \phi_-(y)$; clearly, for the choice (7.22) of ϕ_-, ψ_- satisfies (7.21).

Substituting these values of ϕ_- and ψ_- into $(7.6)'_-$ we get a concrete realization of f_1; we shall denote it by $e_1(z)$:

(7.24)
$$e_1(z) = e_o(z) + \sum c_j^- f_j^+$$

where

(7.24)′
$$e_o(z) = \frac{a^{-iz}}{\sqrt{2}\, iz} \{y^{1/2}(\xi y^{iz}), y^{3/2}\, \partial_y(\xi y^{iz})\}$$

and

(7.24)″
$$c_j^-(z) = E(e_o(z), f_j^-)/\lambda_j^2 \ ;$$

$e_1(z)$ is well defined for $-\lambda < \text{Im } z < 0$. It follows from $(7.8)_-$ that the function ℓ_1 corresponding to the choice (7.22) for ϕ_- is

$$\ell_1(s) = \frac{i}{z}\, \partial_s\, [\xi(ae^{-s})\, e^{-izs}] \ .$$

With this choice for ℓ_1,

$$\ell_2(s) = e^{-izs} - \ell_1 \ .$$

Note that ℓ_2 belongs to L_2 for all z with $\text{Im } z < 0$, and is analytic in z. Therefore $e_2(z)$, the element of $\mathcal{H}$ it represents, depends analytically on z for $\text{Im } z < 0$. Moreover using (7.11) we can carry out the integration involved in computing the $c_j^-(z)$ from (7.24)″; the resulting functions $c_j^-(z)$ are analytic in the whole plane except for poles at the $-i\lambda_j$ for the relevant λ_j's and at $z = 0$.

We now set

(7.25)
$$e(z) = e_1(z) + e_2(z) \ .$$

THEOREM 7.11. *The concrete realization of* $e(z)$ *can be extended into the whole complex plane as a meromorphic function, in the local sense, with poles contained in* $-i\sigma(B'')$ *and the relevant* $\pm i\lambda_j$.

PROOF. We have seen above how $e(z)$ can be realized in the strip (7.18): $-\lambda < \operatorname{Im} z < 0$. We now construct the $\mathfrak{D}'_+$ component of $e(z)$ in this strip z by the second of the two methods discussed earlier in this section. This method is based on the explicit $\mathfrak{D}'_+$ translation representation of $e(z)$, given in Lemma 7.7 as $\mathcal{S}'(z)e^{-izs}$. In this representation the $\mathfrak{D}'_+$ component d'_+ of $e(z)$ is simply

$$d'_+(s) = \begin{cases} \mathcal{S}'(z)e^{-izs} & \text{for} \quad s > 0 \\ 0 & \text{for} \quad s < 0\ . \end{cases}$$

Substituting this into $(7.9)_+$ we get

$$\psi_+(y) = \begin{cases} \dfrac{\mathcal{S}'(z)}{\sqrt{2}}\, a^{iz}\, y^{-iz-1} & \text{for} \quad y > a \\ 0 & \text{for} \quad y < a\ . \end{cases} \tag{7.26}$$

Integrating this we obtain for ϕ_+

$$\phi_+(y) = \begin{cases} \dfrac{i\mathcal{S}'(z)}{z\sqrt{2}}\,(a^{iz}y^{-iz}-1) & \text{for} \quad y > a \\ 0 & \text{for} \quad y < a\ . \end{cases} \tag{7.26$'$}$$

Plugging these formulas for ϕ_+ and ψ_+ into $(7.6)_+$ gives d_+; plugging this d_+ into $(7.5)_+$ gives the coefficients c_j^+. Carrying out the integration involved in $(7.5)_+$ explicitly with the help of (7.11) gives

$$c_j^+(z) = \text{const.}\ \frac{\mathcal{S}'(z)}{(iz+\lambda_j)}\ . \tag{7.26$''$}$$

According to the second definition, the projection $e'_+(z)$ of $e(z)$ onto $\mathcal{D}'_+$ is defined as the element d'_+ defined by formula $(7.6)'_+$ when ϕ_+, ψ_+ and c_j^+ are given by (7.26). These explicit formulas show that $e'_+(z)$ which is equal to the $\mathcal{D}'_+$ component of $e(z)$ in the strip (7.18), can be extended as a *meromorphic function of* z, not as an element of $\mathcal{H}$ but *locally*, that is in any compact subset of the fundamental domain. The poles occur at the points $i\lambda_j$ and at the poles of $\mathcal{S}'(z)$. ■

Similar formulas can be derived for $e'_-(z)$, the $\mathcal{D}'_-$ component of $e(z)$; it too is meromorphic on any compact subset of the fundamental domain with poles at most at the points $-i\lambda_j$.

REMARK. $e'_+(z)$ belongs to $\mathcal{H}$ for $\operatorname{Im} z < 0$ whereas $e'_-(z)$ belongs to $\mathcal{H}$ for $\operatorname{Im} z > 0$.

We have shown already in Lemma 7.10 that $e(z) - e'_+(z) - e'_-(z)$ can be continued as a meromorphic function with poles contained in $-i\sigma(B'')$ and the relevant $-i\lambda_j$. *Thus we see that* $e(z)$ *itself is meromorphic in* z, *in the local sense.* This completes the proof of Theorem 7.11. ■

THEOREM 7.12. $e(z)$ *satisfies the differential equation*

$$(7.27) \qquad (A - iz)e(z) = 0 .$$

PROOF. We first show that (7.27) holds in the strip (7.18). It then follows by the analyticity (in the local sense) of $e(z)$ that (7.27) continues to hold at all z for which $e(z)$ is analytic.

We recall the decomposition (7.25) of $e: e = e_1 + e_2$, where the $\mathcal{D}'_-$ translation representers of e_1 and e_2, denoted by ℓ_1 and ℓ_2, vanish for $s > 0$ and $s < -1$, respectively, and $\ell_1 + \ell_2 = e^{-izs}$. A simple direct calculation based on (7.24) shows that

$$(7.28) \quad (A-iz)e_1 = \frac{a^{-iz}}{\sqrt{2}\, iz}\{y^{1/2}(\xi' y^{iz+1}), y^{3/2}\partial_y(\xi' y^{iz+1})\} + \sum (\lambda_j - iz)c_j^- f_j^+ .$$

Here we have used the relation (3.13):

$$Af_j^+ = \lambda_j f_j^+ ,$$

and the abbreviation ξ' for $\partial_y \xi$. It is clear from (7.28) that

$$k = (A - iz) e_1 \tag{7.29}$$

belongs to $\mathcal{H}$. It follows from Lemma 7.3 that k is represented by

$$-(\partial_s + iz) \ell_1 .$$

On the other hand since ℓ_2 is the $\mathcal{D}'_-$ translation representer of e_2 in $\mathcal{H}$, $(A - iz)e_2$ is represented by

$$-(\partial_s + iz) \ell_2 .$$

Adding these results we see that

$$(A - iz)e = (A - iz)e_1 + (A - iz)e_2$$

is represented by

$$-(\partial_s + iz)(\ell_1 + \ell_2) = -(\partial_s + iz)e^{-izs} = 0 .$$

Thus $(A - iz)e$ is represented by zero and since $(A - iz)e = k + (A - iz)e_2$ belongs to $\mathcal{H}$, it follows that it is zero. This completes the proof of Theorem 7.12. ■

It follows from the differential equation (7.27) that

$$(A - iz)e_2 = -k ,$$

where k is given by (7.29). Since k belongs to $\mathcal{H}$, this equation determines e_2 uniquely for $\operatorname{Im} z < 0$. So we conclude:

LEMMA 7.13. *The differential equation* (7.27) *and the fact that* e *is of the form* $e = e_1 + e_2$, *where* e_1 *is given by* (7.24) *and* e_2 *belongs to* $\mathcal{H}$, *together uniquely determine* e *for* $\operatorname{Im} z < 0$.

We have already shown that the generalized eigenfunctions $e(z)$ which we have constructed in the strip $-\lambda < \operatorname{Im} z < 0$ can be continued analytically, in the local sense, to the whole plane minus the poles of $\mathcal{S}'(z)$ and the relevant $\pm i\lambda_j$. In particular, $e(z)$ can be defined for z real; we shall now show that for z real the generalized eigenfunctions can be used to construct the incoming spectral representation.

THEOREM 7.14. *Denote by* f *data with compact support in* $\mathcal{H}_G$; *denote by* f_c *the projection of* f *into* $\mathcal{H}(=\mathcal{H}_c')$. *Let* ℓ'_- *denote the translation representer of* f_c *with respect to* $\mathcal{D}'_-$, $\tilde{\ell}'_-$ *the corresponding spectral representation. Then*

$$\tilde{\ell}'_-(\sigma) = \frac{1}{\sqrt{2\pi}} E(f, e(\sigma)) \, . \tag{7.30}$$

REMARK 1. Formula (7.30) has appeared already in Section 2 as formula (2.58). In our derivation we follow the argument suggested there.

REMARK 2. The theorem establishes (7.30) for a set of data f dense in $\mathcal{H}_G$; if suitably interpreted the result can be extended to all of $\mathcal{H}_G$.

PROOF. The spectral representation is obtained from the translation representation of f_c by Fourier transformation:

$$\tilde{\ell}'_-(\sigma) = \frac{1}{\sqrt{2\pi}} \int \ell'_-(s) e^{i\sigma s} ds \; ; \tag{7.31}$$

this is obviously a unitary map onto $L_2(\mathbb{R})$.

The projection f_c of f into $\mathcal{H}$ is accomplished by removing from f all eigenfunctions of A which have finite energy. The eigenfunctions corresponding to imaginary eigenvalues of A have vanishing zero Fourier components; the eigenfunctions corresponding to real eigenvalues of A, finite in number and denoted by $f_j^{\pm}$ have, according to (7.11), zero Fourier components of the form

$$\{y^{1/2}\alpha, y^{1/2}\beta\} \tag{7.32}$$

where α and β are linear combinations of the functions $y^{-\lambda_j}$. Since by assumption f has compact support, it follows that $f_c^{(0)}$ also is of form (7.32) for y large. We then deduce from formulas (7.14) and $(7.15)_-$ that the projection f'_- of f_c into $\mathcal{D}'_-$ likewise is of form (7.32) for y large. Substituting this information into formula $(7.8)_-$ we see that for s large enough negative $\ell'_-(s)$ is a linear combination of the $e^{\lambda_j s}$. This enables us to rewrite (7.31) as

$$\tilde{\ell}'_-(\sigma) = \lim_{\tau \uparrow 0} \frac{1}{\sqrt{2\pi}} \int \ell'_-(s)\, e^{i\bar{z}s}\, ds \tag{7.31$'$}$$

where $z = \sigma + i\tau$.

In what follows we shall use the following simple observation:

Since $f - f_c$ is orthogonal to $\mathcal{H}$,

$$E(f, h) = E(f_c, h) \tag{7.33}$$

for all h in $\mathcal{H}$.

LEMMA 7.15. *Let f be an element of $\mathcal{H}_G$ with compact support and g an element of $\mathcal{H}^\lambda$. Denote by f_c the projection of f into $\mathcal{H}$, and by ℓ'_- and g'_- the translation representers of f_c and g in the $\mathcal{D}_-$ translation representation. Then for $\lambda < \lambda_j$*

$$E(f, g) = \int \ell'_-(s)\, \bar{g}'_-(s)\, ds \,. \tag{7.34}$$

PROOF. Denote by g_n that element of $\mathcal{H}$ which is represented by $g_n(s)$ defined by

$$g_n(s) = \begin{cases} g'_-(s) & \text{for} & -n < s \\ 0 & \text{for} & s < -n \ . \end{cases}$$

Since g'_- has finite λ-norm, g_n is in L_2. Since the translation representation is unitary,

$$E(f_c, g_n) = \int f'_-(s) \, \bar{g}_n(s) \, ds \ .$$

Using (7.33) with $h = g_n$ we get

$$(7.34)_n \qquad\qquad E(f, g_n) = \int f'_-(s) \, \bar{g}_n(s) \, ds \ .$$

We claim that as n tends to ∞, the right side of $(7.34)_n$ tends to the right side of (7.34) and the left side of $(7.34)_n$ tends to the left side of (7.34). To see the first part concerning the right sides we use the fact, already established, that as s tends to $-\infty$, $f'_-(s)$ is a linear combination of $e^{\lambda_j s}$, and the assumption that $\lambda < \lambda_j$ and that $g'_- e^{\lambda s}$ is in L_2 on R_-.

To study the left side we note that $g - g_n$ is represented by a function which is zero for $s > -n$, hence in the extended $\mathcal{D}'_-$, and whose λ-norm tends to zero as $n \to \infty$. It follows from the Proposition extended to $\mathcal{H}^\lambda$ that

$$g - g_n = \sum c_j^-(n) f_j^+ \qquad \text{for} \qquad y < ae^n \ ;$$

here $c_j^-(n)$ is given by $(7.5)_-$ with d_- replaced by $g - g_n$. Therefore if n is so large that $f = 0$ for $y > ae^n$, the difference of the left sides of (7.34) and $(7.34)_n$ can be written as

$$E(g - g_n, f) = \sum c_j^-(n) E(f_j^+, f) \ .$$

We have shown already that the $c_j^-(n)$ are bounded functionals in the λ-norm, and since the λ-norm of $g - g_n$ tends to zero as n tends to ∞, so do the $c_j^-(n)$. This completes the proof of Lemma 7.15. ∎

We now apply Lemma 7.15 to $g = e(z)$. This enables us to rewrite (7.31)′ as

$$\tilde{f}'_-(\sigma) = \lim_{\tau \uparrow 0} \frac{1}{\sqrt{2\pi}} \int f'_-(s) e^{i\bar{z}s} ds \tag{7.35}$$

$$= \lim_{\tau \uparrow 0} \frac{1}{\sqrt{2\pi}} E(f, e(z)) \; .$$

We claim that as z tends to σ the right side of (7.35) tends to the right side of (7.30). To see this we decompose $e(z)$ as

$$e(z) = e'_-(z) + e'_+(z) + [e(z) - e'_-(z) - e'_+(z)] \; .$$

According to Lemma 7.10, the expression in brackets on the right is a continuous $\mathcal{H}$-valued function of z near the real axis. On the other hand $e'_+(z)$ is described explicitly by (7.26); in particular it follows from (7.26)″ that $c_j^+(z)$ tends to $c_j^+(\sigma)$ as z tends to σ. Since f is assumed to be zero for y large enough, it follows from (7.26) and (7.26)′ that

$$\lim_{z \to \sigma} E(f, e'_+(z)) = E(f, e'_+(\sigma)) \; ,$$

and similarly for the term containing e'_-. So letting z tend to σ in (7.35) yields the spectral representation formula (7.30), as asserted in Theorem 7.14. ∎

Next we actually compute $e(z)$; for this we shall use the characterization given in Lemma 7.13 of e for $\operatorname{Im} z < 0$ as a solution of

$$(A - iz) e(z) = 0$$

which is of the form $e = e_1 + e_2$, where e_1 is given by (7.24) and e_2 belongs to $\mathcal{H}$. Our starting point is the function h defined as

(7.36) $$h(w) = \text{const.}\,\{y^{1/2+iz},\ izy^{1/2+iz}\}$$

where $w = x + iy$ and

(7.36)′ $$\text{const.} = \frac{a^{-iz}}{\sqrt{2}\,iz}\ .$$

For $y > a+1$, $e_1(w,z) = h(w)$.

The function $h(w)$ is not automorphic since it does not satisfy the relation

(7.37) $$h(w) = h(\gamma w)$$

for all γ of the modular group Γ; (7.37) is however satisfied for all translations since h is independent of x. We therefore omit translations in building the function

(7.38) $$e(w,z) = \sum_{\gamma \epsilon \Gamma_\infty \backslash \Gamma} h(\gamma w)$$

where the summation is over all right cosets of the modular group Γ modulo the integer translations Γ_∞. The expression (7.38) is the celebrated Eisenstein series; since multiplication on the right by any element of Γ permutes the right cosets, it follows that e is an automorphic function of w if the series converges.

It is well known that the series converges for $\operatorname{Im} z < -1/2$; for the sake of completeness we sketch the proof.

With

$$\gamma w = \frac{aw+b}{cw+d}$$

we have, using

(7.39) $$ad - bc = 1\ ,$$

that

$$\operatorname{Im} \gamma w = \frac{y}{|cw+d|^2}\ .$$

Using the definition of (7.36), (7.36)′ of h we deduce from (7.38) that

$$e(z) = e_o \sum \frac{y^{1/2+iz}}{|cw+d|^{1+2iz}} , \tag{7.40}$$

where

$$e_o = \frac{a^{-iz}}{\sqrt{2}}\left\{\frac{1}{iz}, 1\right\} ; \tag{7.40$'$}$$

the summation being over all right cosets. For γ not in the identity coset, $c \neq 0$; multiplying all four coefficients of γ by -1 if necessary we may take $c > 0$. Any other element of Γ in the same coset as γ has the form

$$\frac{aw + b}{cw + d} - n = \frac{(a-nc)w + b - nd}{cw + d} .$$

This shows that every coset contains exactly one transformation for which

$$0 \leq a < c . \tag{7.41}$$

Condition (7.39) implies that c and d are relatively prime:

$$(c, d) = 1 .$$

For such a pair, a and b are uniquely determined by (7.39) and (7.41); so the sum in (7.40) can be written as

$$y^{1/2+iz} + y^{1/2+iz} \sum_{c=1}^{\infty} \sum_{(c,d)=1} \frac{1}{[(cx+d)^2 + c^2y^2]^{1/2+iz}} .$$

We carry out the summation with respect to d by writing

$$d = d_o + mc$$

where

$$0 \leq d_o < c , \quad (d_o, c) = 1 \tag{7.42}$$

and m goes through all integers. We get

$$(7.43)\quad y^{1/2+iz} + y^{1/2+iz} \sum_{c=1}^{\infty} \sum_{\substack{(c_o,d_o)=1 \\ 0<d_o<c}} \sum_{m=-\infty}^{\infty} \frac{1}{[(c(x+m)+d_o)^2+c^2y^2]^{1/2+iz}} .$$

It is well known and not hard to prove that the series on the right converges absolutely, for $\operatorname{Im} z < -1/2$, uniformly on compact subsets of F. We claim that the series converges also in the C-norm defined by

$$(7.44)\qquad C^{1/2}(h) = \left\{ \iint_F \left(h_x^2 + h_y^2 + \frac{h^2}{y^2} \right) dx\, dy \right\}^{1/2} ;$$

it would follow from this that the series represents a function with finite energy.

We sketch the steps in the convergence proof, leaving details to the reader: We use the following simple estimate for the quantity occurring in the denominator, valid for all (x, y) in F:

$$(cx+d)^2 + c^2y^2 \geq \begin{cases} c^2y^2 & \text{if} \quad d < c \\ \dfrac{c^2y^2+d^2}{4} & \text{if} \quad d > c \end{cases} .$$

Using this estimate we get the following inequality:

$$C^{1/2}\left(\frac{y^{1/2+iz}}{|cz+d|^{1+2iz}} \right) \leq \begin{cases} \dfrac{\text{const.}}{c^{1+2\tau}} & \text{for} \quad d < c \\ \dfrac{\text{const.}}{c^{1+2\tau}} \dfrac{1}{q^{1+\tau}} & \text{for} \quad d > c \,; \end{cases}$$

here τ and q are abbreviations:

$$\tau = -\operatorname{Im} z , \quad q = \frac{d}{c} .$$

It follows readily from these estimates that the sum

$$\sum c^{1/2}\left(\frac{y^{1/2+iz}}{|cz+d|^{1+2iz}}\right)$$

converges for $r > 1/2$. This shows that the series (7.38) converges in the energy norm if we exclude the term corresponding to the coset containing the identity.

LEMMA 7.16. a) *The function* $e(z,w)$ *defined by* (7.38) *for* $\operatorname{Im} z < -1/2$ *satisfies the differential equation*

$$(A - iz)e = 0$$

b) $e - e_1$ *has finite energy.*

PROOF. We can verify by a calculation that h, defined by (7.36), formally satisfies $(A - iz)h = 0$. Since the differential operator A is invariant under the action of the group, it follows that also $h(\gamma w)$ is a solution of the same differential equation. Since e is an infinite sum of solutions, it suffices to prove that the series defining e can be differentiated term by term; the estimates needed to prove the convergence of the differentiated series are easy to get, if we allow ourselves the luxury of taking $\operatorname{Im} z$ large enough negative. This completes the proof of part (a).

REMARK. Actually, no additional estimates are necessary; from the convergence of the series defining e, and the fact that $A - iz$ annihilates each term, we conclude that $(A - iz)e = 0$ in the sense of distributions. Since A is an elliptic operator, it follows from elliptic regularity theory that the equation is satisfied pointwise. A further consequence of regularity theory is

COROLLARY 7.17. $e(z,w)$ *is a real analytic function of* w.

We turn now to part (b). We write the series (7.38) defining e as

$$e = h(w) + \sum_{\gamma \in \Gamma_\infty \backslash \Gamma}' h(\gamma w) ,$$

where the $'$ indicates that the summation omits the coset containing the identity. We have already shown that the sum Σ' has finite energy. Comparing formulas (7.34) defining e, and (7.36) defining h we see that since e_o and h are equal for $y > a+1$, $h - e_1$ has finite energy. This proves that $e - e_1$ has finite energy, as asserted in part (b) of the lemma. ■

Combining Lemmas 7.13 and 7.16 leads to

THEOREM 7.18. *The generalized eigenfunction* $e(z)$ *is given by the Eisenstein series* (7.36) *for* $\operatorname{Im} z < -1/2$ *and by its analytic continuation elsewhere.*

Earlier in this section we have described two methods for constructing the $\mathfrak{D}_+'$ component of an element of H^λ. Applying the second method we found that the $\mathfrak{D}_+'$ component of $e(z)$ is of the form $(7.6)_+'$, with ψ_+, ϕ_+ and c_j^+ given by (7.26), (7.26)′ and (7.26)″.

$$\psi_+(y) = \frac{a^{iz}\mathcal{S}'(z)}{\sqrt{2}}\, y^{-iz-1} \qquad \text{for} \qquad y > a . \tag{7.45}$$

We shall now apply the first method for calculating e_+', based on formula $(7.15)_+$; this formula is similar to $(7.6)_+'$, with ψ_+ being given by (7.14):

$$\psi_+ = \frac{1}{2}(\partial_y \alpha - \beta) . \tag{7.46}$$

The quantities α and β are defined by formula (7.13); substituting $e(z)$ for f into that formula gives

$$e^{(0)} = \{y^{1/2}\alpha , y^{3/2}\beta\} . \tag{7.47}$$

Formula (7.40) expresses $e(z)$ as $e_o \Sigma$, where the sum Σ has been rewritten in (7.43). To obtain the zero Fourier coefficient of Σ we integrate (7.43) with respect to x from $-1/2$ to $1/2$; we carry out this process for $\operatorname{Im} z < -1/2$ and then analytically continue the resulting expression to the strip $0 > \operatorname{Im} z > -\lambda$. Note that the sum of integrals indexed by m can be written as a single integral from $-\infty$ to ∞, so that we get

$$\Sigma^{(0)} = y^{1/2+iz} + y^{1/2+iz} \sum_c \sum_{d_o} \int_{-\infty}^{\infty} \frac{dx}{[(cx+d_o)^2 + c^2y^2]^{1/2+iz}} . \tag{7.48}$$

Setting $q = \dfrac{cx+d_o}{cy}$ as new variable of integration, these integrals become:

$$\frac{y}{(cy)^{1+2iz}} \, I \tag{7.49}$$

where

$$I = \int \frac{dq}{(q^2+1)^{1/2+iz}} . \tag{7.50}$$

Since the value of the integral is independent of d_o, summation in (7.48) with respect to d_o can be carried by simply multiplying by $\phi(c)$, the number of d_o between 0 and c which are relatively prime to c. So we get

$$\Sigma^{(0)} = y^{1/2+iz} + I y^{1/2-iz} \sum_{c=1}^{\infty} \frac{\phi(c)}{c^{1+2iz}} . \tag{7.51}$$

In terms of the prime factorization of c:

$$c = \prod p_j^{\alpha_j}$$

ϕ can be expressed as

$$\phi(c) = \prod \left(1 - \frac{1}{p_j}\right) p_j^{\alpha_j} .$$

This suggests writing

$$\sum \frac{\phi(c)}{c^{1+2iz}} = \prod \left[1 + \left(1-\frac{1}{p}\right) \sum_{n=1}^{\infty} \left(\frac{1}{p^{2iz}}\right)^n\right],$$

the product being over all primes. After summing the geometric series the product can be rewritten as

$$\prod\left[1 + \left(1-\frac{1}{p}\right) \frac{p^{-2iz}}{1-p^{-2iz}}\right] = \prod \frac{1-p^{-(1+2iz)}}{1-p^{-2iz}} .$$

Using the product formula for the Riemann ζ function we can write the above expression as

(7.52)
$$\zeta(2iz)/\zeta(2iz+1) = \sum_{1}^{\infty} \frac{\phi(c)}{c^{1+2iz}} .$$

Substituting (7.52) into (7.51) and the result into (7.40) we get the following expression for the zero Fourier coefficient of e:

$$e^{(0)}(z) = \left(y^{1/2+iz} + I y^{1/2-iz} \frac{\zeta(2iz)}{\zeta(1+2iz)}\right) e_o ,$$

where e_o is given by (7.40)′. Comparing this with (7.47) we obtain the following determination for α and β:

$$\alpha = \frac{a^{-iz}}{iz\sqrt{2}}\left[y^{iz} + I \frac{\zeta(2iz)}{\zeta(1+2iz)} y^{-iz}\right]$$

$$\beta = \frac{a^{-iz}}{\sqrt{2}}\left[y^{iz-1} + I \frac{\zeta(2iz)}{\zeta(1+2iz)} y^{-iz-1}\right] .$$

Substituting these into (7.46) we get

(7.53)
$$\psi_+ = -\frac{a^{-iz}}{\sqrt{2}} I \frac{\zeta(2iz)}{\zeta(1+2iz)} y^{-iz-1} .$$

The integral I given by (7.50) can be evaluated in terms of Γ-functions

(7.54) $$I = \frac{\Gamma(1/2)\Gamma(iz)}{\Gamma(1/2+iz)} .$$

Substituting this into (7.53) and comparing the result with our previous determination of ψ_+ in (7.45) we get (cf. Faddeev and Pavlov [6])

THEOREM 7.19. *The scattering matrix is given by the formula*

(7.55) $$\mathcal{S}'(z) = -a^{-2iz} \frac{\Gamma(1/2)\Gamma(iz)\zeta(2iz)}{\Gamma(1/2+iz)\zeta(1+2iz)}$$

where ζ *is the Riemann zeta function.*

Appendix 1 to Section 7: Point spectrum.

Not a great deal is known about the point spectrum of the generator A, even for the modular group. In this appendix we bring together the few facts on this subject of which we are aware.

THEOREM 7.20 (Tanaka [23]). *Let* $N(\lambda)$ *denote the number of purely imaginary eigenvalues of* A *of absolute value less than* λ. *For the modular group*

(7.56) $$\lim_{\lambda\to\infty} \frac{N(\lambda)}{\lambda^2} = \frac{1}{2\pi} \text{ Area } F = \frac{1}{6} .$$

PROOF. Our proof relies on Theorem 8.7 which asserts that

(7.56)′ $$\lim_{\lambda\to\infty} \frac{N(\lambda)+M(\lambda)}{\lambda^2} = \frac{1}{2\pi} \text{ Area } F ,$$

where $M(\lambda)$ denotes the winding number of the scattering matrix:

$$\mathcal{S}'(\sigma) = \frac{\Gamma\left(\frac{1}{2}\right)\Gamma(i\sigma)\zeta(2i\sigma)}{\Gamma\left(i\sigma+\frac{1}{2}\right)\zeta(2i\sigma+1)} a^{-2i\sigma}$$

over the interval $(-\lambda, \lambda)$. The rate of increase of the winding number is given by $-i/2\pi$ times the logarithmic derivative of $\mathcal{S}'(\sigma)$ which is

$$\partial_\sigma \log \mathcal{S}'(\sigma) = -2i \log a + \frac{\partial_\sigma \Gamma(i\sigma)}{\Gamma(i\sigma)} - \frac{\partial_\sigma \Gamma(i\sigma + 1/2)}{\Gamma(i\sigma + 1/2)}$$

$$+ \frac{\partial_\sigma \zeta(2i\sigma)}{\zeta(2i\sigma)} - \frac{\partial_\sigma \zeta(2i\sigma+1)}{\zeta(2i\sigma+1)} .$$

It is known (cf. Titchmarch [24]) that the terms on the right are of order $\log \sigma$ so that

$$2\pi M(\sigma) = |\arg \mathcal{S}'(\sigma) - \arg \mathcal{S}'(-\sigma)| \lesssim \text{const. } \sigma \log \sigma . \tag{7.57}$$

Inserting the estimate (7.57) into (7.56)′ we obtain (7.56). ∎

In the case of the modular group we will eventually prove in Corollary 9.13 that the null space $\mathcal{J}$ is trivial. In the present section we show that $\mathcal{J}$ is at most one dimensional. For the proof we need

LEMMA 7.21. *For data* f *in* $\mathcal{H}_G$ *with* $f_1^{(0)} = 0$ *for* $y > 1$,

$$E(f) \gtrsim \frac{1}{17} G(f) . \tag{7.58}$$

PROOF. The proof depends very much on the shape of the fundamental domain for the modular group as described in (4.2). We begin with the original expression for E:

$$E(f) = \int_F \left\{ |\partial_x f_1|^2 + |\partial_y f_1|^2 - \frac{|f_1|^2}{4y^2} + \frac{|f_2|^2}{y^2} \right\} dx\, dy . \tag{7.59}$$

We shall estimate the negative term on the right by means of the integral of the gradient of f_1. First applying the Parseval relation and using the fact that $f_1^{(0)} = 0$ for $y > 1$, we get

$$(7.60)\qquad 4\pi^2 \iint_{y>1} |f_1|^2\, dx\, dy \le \sum 4\pi^2 k^2 \int_1^\infty |f_1^{(k)}|^2\, dy = \iint_{y>1} |\partial_x f_1|^2 dx\, dy\,.$$

From this it is easy to deduce that

$$(7.61)\qquad \iint_{y>1} \frac{|f_1|^2}{4y^2}\, dx\, dy \le \frac{1}{16\pi^2} \iint_{y>1} |\partial_x f_1|^2\, dx\, dy\ .$$

The corresponding estimate over $F_o = F \cap \{y < 1\}$ is obtained as follows: Write

$$f_1(x,y) = -\int_y^{y+1} \partial_\eta f_1(x,\eta)\, d\eta + f_1(x, y+1)\ .$$

Squaring both sides and applying the Schwarz inequality to the square of the integral on the right, we get

$$|f_1(x,y)|^2 \le 2 \int_y^{y+1} |\partial_y f_1|^2\, dy + 2|f_1(x, y+1)|^2$$

so that

$$\iint_{F_o} |f_1|^2\, dx\, dy \le 2 \iint_F |\partial_y f_1|^2\, dx\, dy + 2 \iint_{y>1} |f_1|^2\, dx\, dy\ .$$

Since $4y^2 > 3$ in F_o we obtain

$$(7.62)\qquad \iint_{F_o} \frac{|f_1|^2}{4y^2}\, dx\, dy \le \frac{2}{3} \iint_F |\partial_y f_1|^2\, dx\, dy + \frac{2}{3} \iint_{y>1} |f_1|^2\, dx\, dy\ .$$

Estimating the second term on the right in (7.62) by (7.60) and combining the resulting inequality with (7.61) we get

$$\iint_F \frac{|f_1|^2}{4y^2}\,dx\,dy \le \frac{2}{3}\iint_F |\partial_y f_1|^2\,dx\,dy + \frac{11}{48\pi^2}\iint_F |\partial_x f_1|^2\,dx\,dy\ . \tag{7.63}$$

Inserting this in (7.59) we see that

$$E(f) \ge \frac{1}{3}\iint_F \left\{|\partial_x f_1|^2 + |\partial_y f_1|^2 + \frac{|f_2|^2}{y^2}\right\} dx\,dy \tag{7.64}$$

$$\ge \frac{1}{8}\iint_{F_o} \frac{|f_1|^2}{y^2}\,dx\,dy = \frac{1}{8}K(f)\ .$$

It follows from this that

$$G(f) = E(f) + 2K(f) \le 17E(f)\ ,$$

as stated in (7.58). ■

Suppose now that f is an eigenfunction of A in $\mathcal{H}_G$:

$$Af = \lambda f\ ,$$

with λ real. Writing this in component form we get

$$f_2 = \lambda f_1 \quad \text{and} \quad Lf_1 = y^2\Delta f_1 + \frac{1}{4}f_1 = \lambda f_2\ .$$

The zero Fourier component of f_1 satisfies the ordinary differential equation:

$$y^2\frac{d^2 f_1^{(0)}}{dy^2} + \frac{1}{4}f_1^{(0)} = \lambda^2 f_1^{(0)} \qquad \text{for} \qquad y \ge 1\ ,$$

whose solutions are of the form:

$$f_1^{(0)} = b_- y^{\frac{1}{2}-|\lambda|} + b_+ y^{\frac{1}{2}+|\lambda|} \qquad \text{if} \quad \lambda \ne 0 \tag{7.65}$$

or

$$(7.65)_0 \qquad f_1^{(0)} = b_1 y^{\frac{1}{2}} + b_2(\log y) y^{1/2} \quad \text{if} \quad \lambda = 0 .$$

It follows from Section 6 that the only real eigenvalues for A correspond to the eigenfunctions in $\mathcal{P}$ or $\mathcal{J}$. In either case $E(f) = 0$ by (3.16) or part (b) of Theorem 5.4. Since $G(f)$ is finite we must have $b_+ = 0 = b_2$. On the other hand we cannot have $b_- = 0$ or $b_1 = 0$; for in this case we would have $f_1^{(0)} = 0$ for $y > 1$ and hence by Lemma 7.21 we would have $E(f) > 0$, which is impossible.

Now if $\mathcal{J}$ were two or more dimensional then some linear combination of eigenfunctions would have $f_1^{(0)} = 0$ for $y > 1$. Likewise if $f_j^{\pm}$ were E-orthogonal to $\mathcal{D}_{\pm}$ and hence not $\pm$ relevant, then by Lemma 6.5, $f_j^{\pm(0)} = \{c\sqrt{y}, 0\}$; since it is also of the form (7.65) we could conclude that $f_j^{\pm(0)} = 0$ for $y > 1$. We have therefore proved:

COROLLARY 7.22. *In the case of the modular group the null space* $\mathcal{J}$ *is at most one dimensional and all of the* λ_j *are relevant.*

We prove in Corollary 9.13 that $\mathcal{J}$ is actually zero dimensional. Moreover it is known that $\mathcal{S}'(z)$ has only one pole in the lower half plane and this is at $z = -i/2$. It therefore follows from Theorem 3.11 and Corollary 7.22 that

COROLLARY 7.23. A *has exactly two real nonzero eigenvalues*; *these are* $\pm 1/2$. *The corresponding eigenfunctions are* $\{1, \pm 1/2\}$.

The above corollary is part of the folk lore in this subject and is attributed to Selberg (see Roelcke [19]).

Appendix 2 to Section 7: How not to prove the Riemann hypothesis.

In this appendix we study a reformulation of the Riemann hypothesis in terms of the semigroup of operators Z'' and the associated scattering operator $\mathcal{S}''$.

According to formula (3.43)

$$\mathcal{S}'' = \mathcal{S}_+ \mathcal{S}' \mathcal{S}_- \ . \tag{7.66}$$

Because of the symmetry inherent in time reversal, $\mathcal{S}_- = \mathcal{S}_+$; and because of the finite dimensionality of $\mathcal{K}_\pm$,

$$\mathcal{S}_- = \mathcal{S}_+ \equiv \mathcal{B}$$

is a Blaschke product. Finally since the spectral multiplicity is one, the scattering matrices are scalars and (7.66) can be rewritten as

$$\mathcal{S}'' = \mathcal{B}^2 \mathcal{S}' \ . \tag{7.67}$$

The Blaschke product $\mathcal{B}$ has as zeros the $-i\lambda_j$, where the λ_j are the relevant positive real eigenvalues of A. It was shown in Corollaries 7.22 and 7.23 for the modular group that there is only one positive real eigenvalue, $\lambda_1 = 1/2$, and it is relevant. Combining this with the determination of $\mathcal{S}'$ obtained in (7.55), we get

$$\mathcal{S}''(z) = -\left(\frac{z+i/2}{z-i/2}\right)^2 \frac{\Gamma\left(\frac{1}{2}\right)\Gamma(iz)\zeta(2iz)}{\Gamma\left(\frac{1}{2}+iz\right)\zeta(1+2iz)} \ . \tag{7.68}$$

The poles of $\mathcal{S}''$ occur at the nontrivial zeros of $\zeta(1+2iz)$ and at $z = i/2$. Therefore the Riemann hypothesis is equivalent to the assertion that $\mathcal{S}''(z)$ has no poles in the half plane $\operatorname{Im} z < 1/4$. Now in Theorem 2.6 we have shown that the poles of the scattering matrix are $-i$ times the spectrum of B'', the generator of the semigroup Z''; the above statement about poles of $\mathcal{S}''$ therefore requires that the eigenvalues μ_j of B'' satisfy the inequality

$$\operatorname{Re} \mu_j \leq -1/4 \ . \tag{7.69}$$

Now if μ_j is an eigenvalue of B'', then

$$\frac{1}{i+\mu_j} e^{\mu_j t}$$

is an eigenvalue of $(B''+i)^{-1}Z''(t)$. The spectral radius cannot exceed the energy norm; that is

$$\left|\frac{1}{i+\mu_j}\right| e^{\operatorname{Re}\mu_j t} \le \|(B''+i)^{-1}Z''(t)\|_E .$$

Therefore the estimate

(7.70) $$\limsup t^{-1} \log \|(B''+i)^{-1}Z''(t)\|_E = -\frac{1}{4}$$

would imply the desired result

(7.71) $$\operatorname{Re}\mu_j \le -\frac{1}{4} .$$

In fact, Faddeev and Pavlov in [6] show that (7.70) is necessary as well as sufficient for (7.71). In this appendix we shall employ the following criterion, similar to but somewhat different from (7.70):

THEOREM 7.23. *The Riemann hypothesis is true if there is a dense set of* f *in* $\mathcal{K}$ *for which*

(7.72) $$\limsup_{t\to\infty} t^{-1} \log \|Z''(t)f\|_E \le -1/4 .$$

PROOF. Denote by E_j the projection onto the eigenspace associated with μ_j:

$$E_j = \frac{1}{2\pi i}\oint (z-B'')dz ,$$

the contour being a small circle about μ_j. Among a dense set of f there is one for which

$$E_j f \neq 0 .$$

Since $E_j f$ is an eigenfunction of Z''

$$Z''(t)E_j f = e^{\mu_j t} E_j f .$$

Also E_j commutes with Z''; so we have

$$\begin{aligned} e^{\mathrm{Re}\mu_j t}\|E_j f\|_E &= \|Z''(t)E_j f\|_E \\ &= \|E_j Z''(t) f\|_E \le \|E_j\|_E \|Z''(t) f\|_E . \end{aligned}$$

Combining this with (7.72) and the fact that $\|E_j f\|_E \neq 0$ we deduce (7.71); this proves the theorem. ∎

The trouble with verifying criterion (7.72), or criterion (7.70) is the difficulty of getting a hold on the space $\mathcal{H}$ which is defined as the orthogonal complement of the proper eigenfunctions of A. Since by Theorem 7.20, A has a substantial point spectrum and since we do not have a precise description of the eigenfunctions of A, we lack a good grip on the space $\mathcal{H}$; this has so far prevented the application of hyperbolic techniques in proving decay results like (7.70) or (7.72).

There is another way of getting at the space $\mathcal{H}$. According to Theorem 6.10, part (iii), the translates $U(t)\mathfrak{D}''_-$ of $\mathfrak{D}''_-$ are dense in $\mathcal{H}$. It will be recalled that the subspaces $\mathfrak{D}_\pm$ depend on an arbitrary parameter $a \geq 1$, corresponding to $\mathfrak{D}^\alpha_+ = U(\alpha)\mathfrak{D}^o_+$ and $\mathfrak{D}^\alpha_- = U(-\alpha)\mathfrak{D}^o_-$; here $\alpha = \log a$. In what follows we take Z'' to be defined in terms of $\mathfrak{D}^{\alpha''}_\pm$ for some $\alpha > 0$ so that

$$\mathcal{K}'' = \mathcal{H} \ominus (\mathfrak{D}^{\alpha''}_- \oplus \mathfrak{D}^{\alpha''}_+) .$$

It is clear from the $\mathfrak{D}''_-$-translation representation of $\mathcal{H}$ that the translates of $\mathfrak{D}''_- \ominus \mathfrak{D}^{\alpha''}_-$ are also dense in $\mathcal{H}$ and it follows from this that

LEMMA 7.24. *The span of the set of elements*

(7.73) $$Z''(t)[\mathfrak{D}''_- \ominus \mathfrak{D}^{\alpha''}_-], \quad t \geq 0$$

is dense in $\mathcal{K}''$.

The purpose of this appendix is to show that using the set (7.73) as the dense set occurring in Theorem 7.23 does not lead to a new approach to the Riemann hypothesis.

An element f of $\mathcal{D}''_-$ is first of all an element of $\mathcal{D}_-$, and hence of the form

$$f = \{y^{1/2}\phi, y^{3/2}\psi\}, \quad \psi = \partial_y \phi , \tag{7.74}$$

where

$$\phi(y) = 0 \quad \text{for} \quad y < 1$$

secondly f is orthogonal to the eigenfunctions of A with real nonzero eigenvalues. In the present case there are only two such eigenfunctions,

$$f_1^{\pm} = \{1, \pm 1/2\} .$$

Orthogonality of f to these in the energy norm is equivalent with

$$\int_1^\infty \frac{\phi(y)}{y^{3/2}}\, dy = 0 . \tag{7.75}$$

The space $\mathcal{D}^{a''}_-$ consists of similar functions, except that $\phi(y)$ is required to vanish for $y < a$. It follows that $\mathcal{D}''_- \ominus \mathcal{D}^{a''}_-$ consists of all f of form (7.74) where ϕ satisfies (7.75) and is of the form

$$\phi(y) = \begin{cases} 0 & \text{for} \quad y < 1 \\ \dfrac{\text{const.}}{y^{3/2}} & \text{for} \quad a < y \end{cases} \tag{7.76}$$

Given f in $\mathcal{D}''_- \ominus \mathcal{D}^{a''}_-$, to compute $Z''(t)f$ we first compute $U(t)f$. To do this we begin by constructing a solution of the wave equation (6.1) using formula $(6.9)_-$:

$$u_o(w, t) = y^{1/2}\phi(ye^t) . \tag{7.77}$$

This solution is automorphic for $t \le 0$, has initial data f given by (7.76) but is not automorphic for $t > 0$. To make it so we sum over the right cosets $\Gamma_\infty \backslash \Gamma$:

$$u(w,t) = \sum_{\Gamma_\infty \backslash \Gamma} u_0(\gamma w, t) .$$

We use the same description of the right cosets which we did at the end of Section 7 in evaluating the zero component of the Eisenstein function, see e.g. the sum (7.43). We get

$$(7.78) \qquad u(w,t) = u_0(w,t) + \sum_{\substack{(c,d)=1 \\ 0 \le d < c}} \sum_{m=-\infty}^{\infty} \left[\frac{y}{(c(x+m)+d)^2 + c^2y^2}\right] \cdot$$

$$\cdot\, \phi\left(\frac{ye^t}{(c(x+m)+d)^2 + c^2y^2}\right) .$$

REMARK. In any bounded time interval $t \le T$ there are only a finite number of nonzero terms in the above sum.

To compute $Z''(t)$ we have to remove the $\mathfrak{D}^{a''}_{\pm}$ components of $U(t)$. We shall remove more, and show the difficulty one runs into when trying to derive an estimate of the form (7.72) for one of the remaining components of $Z''f$. The component in question is the first Fourier component of u:

$$u^{(1)} = \int_{-1/2}^{1/2} e^{-2\pi ix} u(x,y,t)\,dx .$$

We substitute the series (7.78) for u into the above integral; the summation with respect to m can be replaced by extending the integration over the whole x-axis. We get

$$(7.79)\qquad u^{(1)}(y,t) = \int_{-\infty}^{\infty} \sum_c \sum_d \left[\frac{y}{(cx+d)^2 + c^2y^2}\right]^{1/2}$$

$$\cdot\ e^{-2\pi i x}\,\phi\left(\frac{ye^t}{(cx+d)^2 + c^2y^2}\right) dx\ .$$

For c, d fixed we introduce r as new variable of integration:

$$\frac{1}{y}\left(x + \frac{d}{c}\right) = r\ .$$

We get

$$(7.80)\qquad u^{(1)}(y,t) = y^{1/2} \sum_{c,d} \frac{e^{2\pi i \frac{d}{c}}}{c} \int_{-\infty}^{\infty} \frac{e^{-2\pi i y r}}{(r^2+1)^{1/2}}\, \phi\left(\frac{e^t}{yc^2(r^2+1)}\right) dr$$

$$= y^{1/2} \sum_{c=1}^{\infty} \frac{\mu(c)}{c}\, I(c,y,t)\ ,$$

where

$$(7.81)_1\qquad \mu(c) = \sum_{\substack{(c,d)=1 \\ 0 \le d < c}} e^{2\pi i d/c}$$

and

$$(7.81)_2\qquad I(c,y,t) = \int_{-\infty}^{\infty} \frac{e^{-2\pi i y r}}{(r^2+1)^{1/2}}\, \phi\left(\frac{e^t}{yc^2(r^2+1)}\right) dr\ .$$

Finally from (7.80) we see that the 1-Fourier component of u_t is

$$(7.80)'\qquad u_t^{(1)}(y,t) = y^{1/2} \sum_{c=1}^{\infty} \frac{\mu(c)}{c}\, \partial_t\, I(c,y,t)\ .$$

Denoting the $L_2(F_1)$ norm by

$$\|g\|^{F_1} = \left[\iint_{F_1} |g|^2 \frac{dx\,dy}{y^2}\right]^{1/2}$$

it is clear that

$$\|u_t'(t)\|^{F_1} \le \|Z''(t)f\|_E .$$

Hence if the inequality (7.72) were true, then we would have

$$\limsup_{t\to\infty} t^{-1} \log \|u_t^{(1)}(t)\|^{F_1} \le -1/4 ;$$

and the Laplace transform of $u_t^{(1)}$,

$$\int_0^\infty e^{-zt} u_t^{(1)}(y,t)\,dt = \hat{u}_t^{(1)}(y,z) \tag{7.82}$$

would be analytic in the half-plane $\operatorname{Re} z > -\frac{1}{4}$.

We now compute this Laplace transform. Substituting (7.80)′ into (7.82) we get

$$\hat{u}_t^{(1)}(y,z) = y^{1/2} \sum_{c=1}^{\infty} \frac{\mu(c)}{c} \int_0^\infty e^{-zt}\, \partial_t I(c,y,t)\,dt . \tag{7.83}$$

We use the definition $(7.81)_2$ of I in the above integral, and introduce

$$q = \frac{e^t}{yc^2(r^2+1)}$$

as new variable of integration so that

$$e^{-zt} = [qyc^2(r^2+1)]^{-z} \quad \text{and} \quad \frac{dq}{q} = dt ;$$

we then get

(7.83)′ $$\int_0^\infty e^{-zt}\partial_t I(c,y,t)\,dt = \frac{1}{(yc^2)^z}\int_{-\infty}^{\infty}\frac{e^{-2\pi iyr}}{(r^2+1)^{1/2+z}}\int^{\infty}\frac{\phi'(q)}{q^z}\,dq\,dr\ .$$

The lower limit of the q integration is $[yc^2(r^2+1)]^{-1}$; since $\phi(q) = 0$ for $q < 1$, for $y > 1$ the lower limit can be taken to be $= 1$. This makes the right side of (7.83)′ a product where the last two factors are independent of c:

(7.83)″ $$\frac{1}{y^z c^{2z}}\ J(y,z)\Phi(z)$$

where

(7.84)$_1$ $$J(y,z) = \int_{-\infty}^{\infty}\frac{e^{-2\pi iyr}}{(r^2+1)^{1/2+z}}\ dr$$

and

(7.84)$_2$ $$\Phi(z) = \int_1^{\infty}\frac{\phi'(q)}{q^z}\ dz\ .$$

Substituting all this back into (7.83) we get

(7.85) $$\hat{u}_t^{(1)}(y,z) = y^{1/2-z}\sum_{c=1}^{\infty}\frac{\mu(c)}{c^{1+2z}}\ J(y,z)\Phi(z)\ .$$

The finite sum (7.81)$_1$ defining μ can be evaluated explicitly:

$$\mu(c) = \begin{cases} 1 & \text{for } c = 1, \\ (-1)^m & \text{if } c \text{ is the product of } m \text{ distinct primes}, \\ 0 & \text{otherwise}. \end{cases}$$

μ is called the Möbius function. The Dirichlet series appearing in (7.85) is well known to be (see Titchmarsh [p. 3, 24].

(7.86) $$\sum \frac{\mu(c)}{c^{1+2z}} = \frac{1}{\zeta(1+2z)}\ .$$

Substituting this into (7.85) we get

$$\hat{u}_t^{(1)}(y,z) = y^{1/2-z}\,\frac{J(y,z)\Phi(z)}{\zeta(1+2z)}\,. \tag{7.87}$$

If we integrate by parts repeatedly in the integral $(7.84)_1$ defining $J(y,z)$ we find that for $y \neq 0$, J is an entire analytic function of z. Since by (7.76), $\phi(q) = \frac{\text{const.}}{q^{3/2}}$ for $q > a$, we conclude that $\Phi(z)$ defined by $(7.84)_2$ is analytic for $\operatorname{Re} z > -5/2$. On the other hand the denominator in formula (7.87) for $\hat{u}_t^{(1)}(y,z)$ vanishes when $2z+1$ is a zero of the ζ function. Therefore if we knew that the Laplace transform of $u_t^{(1)}$ were analytic for $\operatorname{Re} z > -1/4$, we could deduce directly that the ζ-function has no zeros to the right of the critical line.

In fact, any attempt to prove directly on the basis of formula (7.80) that $u^{(1)}(y,t)$ decays exponentially in t necessarily runs into the same difficulty as one encounters in trying to prove directly that the Dirichlet series (7.86) for the reciprocal of the ζ-function converges up to the critical line. We believe that the only way scattering theory can be used to approach the Riemann hypothesis is through a characterization of the space $\mathcal{H}$ which does not involve $\mathcal{D}''_-$.

§8. THE GENERAL CASE

The foregoing theory can easily be adapted to furnish a scattering theory for the wave equation acting on functions automorphic with respect to a general discrete subgroup Γ of G whose fundamental domain has a finite number of cusps, in other words, a noncompact domain of finite area.

For such a discrete subgroup one can construct a fundamental domain F with the following properties:

a) F lies in a strip: $-X < x < X$ for some X.

b) $F = \bigcup_{j=0}^{n} F'_j$, where $F'_0 = F_0$ is compact with piecewise smooth boundary and F'_j, $j = 1, 2, \cdots, n$, is the image of the half strip

$$F_1 = \{-1/2 < x < 1/2,\ y > a\} \tag{8.1}$$

under the mapping: $w \to g_j w$, for some g_j in G and some a sufficiently large.

The domains F'_j are neighborhoods of the n nonequivalent cusps. To each function f defined on F we associate $n+1$ *component functions* f_j defined by

$$\begin{cases} f_0(w) = f(w) & \text{for } w \text{ in } F_0 \\ f_j(w) = f(g_j w) & \text{for } w \text{ in } F_1 \text{ and } j = 1, 2, \cdots, n\ . \end{cases} \tag{8.2}$$

It is clear that f and its components determine each other uniquely.

The sides of F split into pairs determined by the fact that the two sides of a pair map into each other under some transformation (and its inverse) in Γ. This same transformation also relates pairs of points p and $\bar{p}$ within each of the two sides of the pair. If these corresponding points are identified then F forms a manifold; however we prefer to work with the automorphic functions defined on F. In this case the domain of the Laplace-Beltrami operator L_0 is restricted by the boundary conditions:

$$f(p) = f(\bar{p}) \quad \text{and} \quad f_n(p) = -f_n(\bar{p}) , \tag{8.3}$$

where again p and $\bar{p}$ are corresponding boundary points and f_n denotes the outward normal derivative of f.

Starting with $L_0 = y^2\Delta$ defined on all C^2 functions with bounded support and satisfying the boundary conditions (8.3), we proceed as in Section 4 to construct the self-adjoint Friedrichs' extension. In fact all of the results of Section 4 remain valid for the many cusp fundamental domain; the only modifications required in the arguments involve rewriting the equations so as to take into account the various components of the functions. For instance equation (4.25) with

$$L = y^2\Delta + \frac{1}{4} \tag{8.4}$$

becomes

$$-(u, Lu) = \iint_{F_0} \left\{ (\partial_x u_0)^2 + (\partial_y u_0)^2 - \frac{u_0^2}{4y^2} \right\} dx\,dy \tag{8.5}$$

$$+ \sum_{j=1}^{n} \iint_{F_1} \left\{ (\partial_x u_j)^2 + y\left[\partial_y\left(\frac{u_j}{\sqrt{y}}\right)\right]^2 \right\} dx\,dy - \sum_{j=1}^{n} \int_{y=a} \frac{u_j^2}{2y}\, dx .$$

Since the proof of Lemma 4.2 involves only an integration by parts in the cusp neighborhood for $y < a$, it is clear that by choosing a sufficiently large we can prove the following analogue of the estimate (4.26):

$$\sum_{j=1}^{n} \int_{y=a} \frac{u_j^2}{2y}\, dx \leq \iint_{F_0} \left\{ \frac{3}{2}\,\frac{u_0^2}{y^2} + \frac{1}{2}\,(\partial_y u_0)^2 \right\} dx\,dy . \tag{8.6}$$

No further changes are needed in Section 4.

Likewise only perfunctory modifications are required in Section 5. Making use of (8.5) the expression for the energy form

$$E = \|u_t\|^2 - (u, Lu) \tag{8.7}$$

can be rewritten as

$$E = \iint_{F_0} \left\{ (\partial_x u_0)^2 + (\partial_y u_0)^2 - \frac{u_0^2}{4y^2} + \frac{(\partial_t u_0)^2}{y^2} \right\} dx\,dy$$

$$+ \sum_{j=1}^{n} \iint_{F_1} \left\{ (\partial_x u_j)^2 + y \left[\partial_y \left(\frac{u_j}{\sqrt{y}} \right) \right]^2 + \frac{(\partial_t u_j)^2}{y^2} \right\} dx\,dy$$

$$- \sum_{j=1}^{n} \int_{y=a} \frac{u_j^2}{2y}\, dx\,.$$

For data $u = \{u_1, u_2\}$ we define

$$K(u) = K(u_1) = \iint_{F_0} \frac{(u_1)_0^2}{y^2}\, dx\,dy \tag{8.8}$$

and

$$G(u) = E(u) + 2K(u) \tag{8.9}$$

as before. Then, aside from the obvious changes required by the presence of several cusp components, the rest of the section goes over as is.

Incoming and outgoing solutions exist in each cusp neighborhood; in terms of the component functions these solutions are of the form:

$$u_j(w,t) = \begin{cases} y^{1/2}\phi_j(ye^t) & \text{for } t \le 0 \text{ and } j \ne 0\,, \\ 0 & \text{for } t \le 0 \text{ and } j = 0\,; \end{cases} \tag{8.10$_-$}$$

$$(8.10)_+ \qquad u_j(w,t) = \begin{cases} y^{1/2}\phi_j(ye^{-t}) & \text{for } t \geq 0 \text{ and } j \neq 0, \\ 0 & \text{for } t \geq 0 \text{ and } j = 0; \end{cases}$$

here the ϕ_j vanish for $y \leq a$. In analogy with (6.10) we define the incoming and outgoing subspaces to be the closure in $\mathcal{H}_G$ of the initial data for the above incoming and outgoing solutions:

$$\mathcal{D}_- = \text{closure}\,\{f;\, f_0 = 0,\, (f_1)_j = y^{1/2}\phi_j(y),\, (f_2)_j = y^{3/2}\phi_j'(y) \text{ for } j \neq 0\},$$

(8.11)

$$\mathcal{D}_+ = \text{closure}\,\{f;\, f_0 = 0,\, (f_1)_j = y^{1/2}\phi_j(y),\, (f_2)_j = -y^{3/2}\phi_j'(y) \text{ for } j \neq 0\},$$

where the functions ϕ_j are of class C^1, of bounded support and vanish for $y < a$.

As before these data can be used to obtain translation representations for $\mathcal{D}_-$ and $\mathcal{D}_+$; they are, respectively,

$$(8.12)_- \qquad \sqrt{2}\, ae^{-s}\phi_j'(ae^{-s}) \qquad j = 1, 2, \cdots, n,$$

and

$$(8.12)_+ \qquad \sqrt{2}\, ae^{s}\phi_j'(ae^{s}) \qquad j = 1, 2, \cdots, n.$$

It is clear that $\mathcal{D}_-$ maps isometrically onto $L_2(\mathbf{R}_-, \mathfrak{N})$ and $\mathcal{D}_+$ onto $L_2(\mathbf{R}_+, \mathfrak{N})$ where the auxiliary space $\mathfrak{N}$ is now n-dimensional. We can therefore anticipate that the spectrum of A on $\mathcal{H}_c'$ will be of uniform multiplicity n.

Lemmas 6.3 and 6.5 remain valid in each of the cusp neighborhoods. The changes required in the proof of Theorem 6.6 are trivial: Part (a), dealing with a compact subset of F, holds as is whereas the argument of part (b) can be applied separately to each of the cusp neighborhoods. The rest of Section 6 can be taken over verbatim.

The changes required in Section 7 are somewhat more substantive but still harmless. We shall mention the main lines of the argument, inserting proofs when needed.

As in (8.11) each cusp neighborhood furnishes a different component of the data in $\mathfrak{D}_-$ and $\mathfrak{D}_+$, and each element d'_- $[d'_+]$ of $\mathfrak{D}'_-$ $[\mathfrak{D}'_+]$ is obtained from some d_- $[d_+]$ in $\mathfrak{D}_-$ $[\mathfrak{D}_+]$ as in (7.4). Thus for

$$d_{\pm j} = \begin{cases} \{y^{1/2}\phi_{\pm j}(y), \mp y^{3/2}\psi_{\pm j}(y)\} & \text{for } j \neq 0, \\ 0 & \text{for } j = 0, \end{cases} \tag{8.13}$$

where

$$\psi_j(y) = \partial_y \phi_j(y)$$

and the functions $\phi_j(y) = 0$ for $y < a$, the corresponding cusp components of $d'_\pm$ are

$$d'_{\pm j} = \begin{cases} \{y^{1/2}\phi_{\pm j}(y), \mp y^{3/2}\psi_{\pm j}(y)\} + \sum c_k^{\pm}(f_k^{\mp})_j & \text{for } j \neq 0, \\ \sum c_k^{\pm}(f_k^{\mp})_0 & \text{for } j = 0, \end{cases} \tag{8.13$'$}$$

the constants $c_k^{\pm}$ being determined by

$$c_k^{\pm} = E(d_\pm, f_k^{\pm})/\lambda_k^2 . \tag{8.14}$$

The $\mathfrak{D}'_\pm$ translation representer components of an element of the form (8.13)′ are

$$d'_{\pm j}(s) = \sqrt{2}\, ae^{\pm s}\psi_{\pm j}(ae^{\pm s}), \qquad j = 1, 2, \cdots, n . \tag{8.15}$$

This relation can be inverted to express ψ in terms of d':

$$\psi_{-j}(y) = \frac{1}{\sqrt{2y}}\, d'_{-j}(\alpha - \log y) \tag{8.16$_-$}$$

$$(8.16)_+ \qquad \psi_{+j}(y) = \frac{1}{\sqrt{2y}}\, d'_{+j}(\log y - \alpha)$$

where $\alpha = \log a$. Integrating ψ we can recover ϕ to obtain d by (8.13), c_k by (8.14) and then d' by (8.13)′.

This correspondence can be extended to any function d'_- supported on $\mathbf{R}_-$ having finite λ-norm; we denote by $\mathcal{H}^\lambda$ the set of data so constructed. Likewise the E-orthogonal projection of f onto $\mathcal{D}'_+$ given by $(7.15)_+$ can be extended to any f in $\mathcal{H}^\lambda$ as in Lemma 7.5. Finally we can define translation representations with respect to the subspaces $\mathcal{D}''_-$, $\mathcal{D}''_+$ and $\mathcal{D}'_+$ for any element of $\mathcal{H}^\lambda$ by applying successively the operators S_-^{-1}, S'' and S_+^{-1} to the $\mathcal{D}'_-$ translation representation of this element. This requires only that these operators be continuous in the λ-metric; this follows from Theorem 2.13 in the case of S'' and from Theorem 2.18 in the cases S_-^{-1} and S_+^{-1}.

From the $\mathcal{D}'_+$ outgoing translation representation ℓ'_+ of f given as above by

$$(8.17) \qquad \ell'_+ = S'\ell'_- = S_+^{-1} S'' S_-^{-1} \ell'_-$$

we construct the $\mathcal{D}'_+$ component d'_+ of f by setting d'_+ equal to the restriction of ℓ'_+ to $\mathbf{R}_+$. One proves as in Lemma 7.6 that this method of obtaining d'_+ and that of the direct method of Lemmas 7.4 and 7.5 give the same answer.

Next we construct a basis of n generalized eigenvectors $e^k(z, w)$ whose $\mathcal{D}'_-$ translation representers ℓ^k are given in component form by

$$(8.18) \qquad \ell_j^k(z, s) = \{e^{-izs}\delta_{jk};\ j = 1, 2, \cdots, n\}\ .$$

these functions will be of finite λ-norm if z lies in the strip: $-\lambda < \operatorname{Im} z < 0$. We denote the concrete realization of these generalized eigenvectors for such z by $e^k(z)$. As in Lemma 7.7 the translation representation of $e^k(z)$ with respect to $\mathcal{D}'_-$, $\mathcal{D}''_-$, $\mathcal{D}''_+$ and $\mathcal{D}'_+$ are, respectively,

(8.19) $$\rho^k(z), \mathcal{C}_-(z)\rho^k(z), \mathcal{S}''(z)\mathcal{C}_-(z)\rho^k(z), \mathcal{S}'(z)\rho^k(z) .$$

Proceeding as in Section 7 we decompose ρ^k as in (7.12):

$$\rho^k = \rho_1^k + \rho_2^k$$

where $\rho_2^k(s) = 0$ for $s < -1$. ρ_1^k can be realized as

(8.20) $$f_1^k = f_0^k + \sum c_\ell^{k-} f_\ell^+$$

where the jth cusp component of f_0^k is given by

(8.20)′ $$(f_0^k)_j = \begin{cases} \dfrac{a^{-iz}}{\sqrt{2}\,iz}\,\delta_{jk}\{y^{1/2}(\xi y^{iz}), y^{3/2}\,\partial_y(\xi y^{iz})\} & \text{for } j \neq 0 \\ 0 & \text{for } j = 0 \end{cases}$$

and

$$c_\ell^{k-} = E(f_0^k, f_\ell^-)/\lambda_\ell^2 ,$$

ξ being defined as in (7.23). Note that this can be continued analytically to all z with $\operatorname{Im} z < 0$ and that ρ_2^k is in L_2 for all such z.

The proofs of Theorems 7.11 and 7.12 now go over with obvious modifications and we can assert that the concrete realization of the $e^k(z)$ can be extended into the whole complex plane as a meromorphic function, in the local sense, with poles contained in $-i\sigma(B'')$ and the relevant $\pm i\lambda_\ell$; and further that

(8.21) $$(A - iz)\,e^k(z) = 0 .$$

Likewise the analogue of Lemma 7.13 holds.

The generalized eigenfunctions can be used to construct the incoming spectral representation.

THEOREM 8.1. *Denote by* f *data with compact support in* $\mathcal{H}_G$ *and denote by* f_c *the* E*-orthogonal projection of* f *into* $\mathcal{H}_c'$. *Finally let* $\tilde{f}_-'$ *denote the spectral representation of* f_c *with respect to* $\mathcal{D}_-'$. *Then component-wise*

$$(8.22) \qquad (\tilde{f}_-')_k(\sigma) = \frac{1}{\sqrt{2\pi}} E(f, e^k(\sigma)), \qquad k = 1, 2, \cdots, n .$$

The proof of this theorem is essentially the same as that of Theorem 7.14.

The generalized eigenfunction $e^k(z)$ can be constructed fairly explicitly by means of the Eisenstein series more or less as before. In this case it is convenient to replace F by $F^k = g_k^{-1}F$ and Γ by the conjugate group

$$\Gamma^k = g_k^{-1}\Gamma g_k .$$

This has the effect of transforming the kth cusp neighborhood into F_1 and the stabilizer of the kth cusp into the integer translations Γ_∞. As before our starting point is the function h, defined in formula (7.36), which is formally an eigenfunction of A but not automorphic. Taking into account the fact that h is invariant under translations, an automorphic eigenfunction can be obtained from h by summing over the right cosets $\Gamma_\infty \backslash \Gamma^k$:

$$(8.23) \qquad e_k(z, w) \equiv \sum_{\gamma \epsilon \Gamma_\infty \backslash \Gamma^k} h(\gamma w) .$$

Once it is shown that the Eisenstein series (8.23) less h converges in the G-norm, then one can prove as in Lemma 7.16 that the $e_k(z, w)$ so constructed is indeed a generalized eigenfunction, that is that

$$(A - iz) e_k(z, w) = 0 .$$

LEMMA 8.2. *For* $\text{Im}\, z \leq -4$ *the sum* $e_k - h$ *converges in the* G*-norm over* F^k.

PROOF. We choose as representatives of the above right cosets transformations γ for which γF^k lies within the strip $\{|x| < 1\}$ and we order the so chosen γ's : γ_0 = identity, $\gamma_1, \gamma_2, \cdots$. With

$$\gamma_j w = \frac{a_j w + b_j}{c_j w + d_j}$$

and

$$a_j d_j - b_j c_j = 1 ,$$

we show first of all that

$$|c_j| \to \infty . \tag{8.24}$$

In fact it is clear from our choice of γ_j's that $\gamma_j \infty = \infty$ only for $j = 0$ and further that infinity is no limit point of $\gamma_j w$ for any w in F^k. Consequently $c_j \neq 0$ for $j > 0$ and the two sequences of real numbers

$$\gamma_j(\infty) = a_j/c_j \quad \text{and} \quad \operatorname{Im} \gamma_j(ai) = a/[(c_j a)^2 + d_j^2] \tag{8.25}$$

are bounded for $j > 0$. Suppose some subsequence of the $|c_j|$ were bounded. Then because of (8.25) there is a sub-subsequence, which we still denote by γ_j, for which

$$c_j \to c^0,\ a_j \to a^0,\ d_j \to d^0 \quad \text{and} \quad b_j = c_j^{-1}(a_j d_j - 1) \to b^0 ;$$

however this is impossible for a discrete subgroup such as Γ^k.

This being so there is a standard procedure for proving that

$$\sum_{j>0} |c_j|^{-r} < \infty \tag{8.26}$$

for $r \geq 4$; see, for instance, J. Lehner [p. 159, 17].

It is clear that the G-norm of the individual terms in (8.23) is majorized by the C-form (see (4.9)) of the first component of the $h(\gamma w) = \{h_1(\gamma w), h_2(\gamma w)\}$.

Making use of the invariance of the Dirichlet form we get

$$C(h_1(\gamma w)) = \iint_{F^k} \left\{ |\partial_x h_1(\gamma w)|^2 + |\partial_y h_1(\gamma w)|^2 + \frac{|h_1(\gamma w)|^2}{y^2} \right\} dx\,dy$$

$$= \iint_{\gamma F^k} \left\{ |\partial_x h_1(w)|^2 + |\partial_y h_1(w)|^2 + \frac{|h_1(w)|^2}{y^2} \right\} dx\,dy$$

$$= c(z) \iint_{\gamma F^k} \frac{dx\,dy}{y^{1+2\,\mathrm{Im}\,z}} ,$$

where $c(z) = \left(1 + |\frac{1}{2} + iz|^2\right) \frac{a^{2\,\mathrm{Im}\,z}}{2|z|^2}$.

Denoting the isometric circle of the transformation γ by

$$\mathcal{I}(\gamma) : |w + d/c| = 1/|c| ,$$

it is easy to show for $c \neq 0$ that γ maps the exterior of $\mathcal{I}(\gamma)$ onto the interior of $\mathcal{I}(\gamma^{-1})$ (see, for instance, Lehner [p. 75, 17]). Hence because of (8.24) it is clear that for j sufficiently large, $\gamma_j F^k$ is contained in $\mathcal{I}(\gamma_j^{-1})$, in which case

$$C(h_1(\gamma_j w)) \leq c(z) \int_0^{1/|c_j|} \frac{dy}{y^{1+2\,\mathrm{Im}\,z}} = c'(z)|c_j|^{2\,\mathrm{Im}\,z}, \; c'(z) = -\frac{c(z)}{2\,\mathrm{Im}\,z} .$$

It now follows by (8.25) that for $\mathrm{Im}\,z \leq -4$ the sum (8.23) converges in the G-norm over F^k if we exclude the γ_0 term. This completes the proof of Lemma 8.2. ∎

REMARK 1. It is evident from the proof of Lemma 8.2 that $c'(z)$ is bounded on compact subsets of the half-plane $\mathrm{Im}\,z \leq -4$. Consequently the convergence of (8.23) is uniform in z on such subsets and hence $e_k(z, w)$ is analytic in z in the half-plane.

Since $e_k - h$ is of finite energy on F^k and since h is of finite energy on $F^k \backslash F_1$, we see that the zero Fourier components of the various cusp neighborhoods must also be of finite energy and hence outgoing except for the kth cusp where there is also an incoming part due to $h(w)$. The zero Fourier component of the jth cusp component of e_k is therefore of the form

$$(8.27) \qquad (e_k)_j^{(0)} = \frac{a^{-iz}}{\sqrt{2}\, iz}\,(\delta_{kj}\, y^{\frac{1}{2}+iz} + s_{kj}(z)\, y^{\frac{1}{2}-iz})\, \{1, iz\}\ .$$

Comparing the formulas (8.20) defining f_1^k and (8.27) describing the incoming and outgoing parts of the Eisenstein series e_k we see that $e_k - f_1^k$ has finite energy. It now follows by Lemma 7.13 that the previously defined generalized eigenfunction e^k and the Eisenstein series e_k coincide for $\operatorname{Im} z \leq -4$. The Eisenstein series therefore have meromorphic continuations to the whole complex plane.

The Eisenstein series can now be used to obtain an explicit formula for the scattering matrix. Recall that the scattering operator is determined in general by the $\mathcal{D}'_+$ projection of the generalized eigenfunctions e^k. This projection is of the form (8.13)′ where ψ_{+j}, ϕ_{+j} and c_k^+ are given by the analogue of (7.26); in particular

$$(8.28) \qquad (\psi_+^k)_j = \frac{a^{iz}}{\sqrt{2}}\, \mathcal{S}'_{jk}(z)\, y^{-iz-1} \qquad \text{for } y > a\ .$$

The $\mathcal{D}'_+$ projection of $e_k = e^k$ can also be computed directly from (8.27) as in (7.53), giving

$$(8.29) \qquad (\psi_+^k)_j = -\frac{a^{-iz}}{\sqrt{2}}\, s_{kj}(z)\, y^{-iz-1} \qquad \text{for } y > a\ .$$

Comparing (8.28) and (8.29) we have

THEOREM 8.3. *The scattering matrix* $\mathcal{S}'(z)$ *is determined by the Eisenstein series for* $\operatorname{Im} z \leq -4$ *as*

(8.30) $$\mathcal{S}'_{jk}(z) = -a^{-2iz} s_{kj}(z) ,$$

where s_{kj} *is defined by* (8.27).

We have now extended all of the material in the previous sections to the many cusp case. We conclude Section 8 with two further results; the first is a Rellich-like uniqueness theorem and the second concerns the asymptotic distribution of the eigenvalues of A.

THEOREM 8.4. *Suppose that* f *is a generalized eigenfunction of* A:

$$Af = izf ,$$

that f *is outgoing in the sense of being* E-*orthogonal to all elements of* $\mathcal{D}_-$ *with compact support and that* $f - f^{(0)}$ *is of finite* G-*norm. If* $\operatorname{Im} z \leq 0$, $z \neq 0$ *and* f *is not a proper eigenfunction of* A, *then* $f \equiv 0$.

REMARK. More precisely we require of each cusp component of f that f_0 and $f_j - \xi f_j^{(0)}$ for $j \neq 0$ be of finite G-norm; here ξ denotes a C^∞ function such that

$$\xi(y) = \begin{cases} 1 & \text{for} \quad y > a+1 \\ 0 & \text{for} \quad y < a . \end{cases}$$

PROOF. The relation $Af = izf$ can be rewritten in component form as

(8.31) $$f_2 = iz f_1$$
$$y^2 \Delta f_1 + \frac{1}{4} f_1 = iz f_2 = -z^2 f_1 .$$

In the jth cusp neighborhood the zero Fourier component of the solution to this system is of the form

$$(f_1)_j^{(0)} = y^{1/2}(b_j y^{iz} + c_j y^{-iz})$$
$$(f_2)_j^{(0)} = iz y^{3/2}(b_j y^{iz} + c_j y^{-iz})' .$$

According to Corollary 6.4, f will be outgoing only if all of the $b_j = 0$. If all of the $c_j = 0$ or if $\operatorname{Im} z < 0$, then f is of finite G-norm and hence a proper eigenfunction. It remains to consider the case when $z = \sigma$ is real nonzero, and some c_j is different from zero.

We choose

$$\omega = 1 - \xi = \begin{cases} 0 & \text{for} \quad y > a+1 \\ 1 & \text{for} \quad y < a \end{cases}$$

and set

$$\omega_t(y) \equiv \omega(ye^{-t}) .$$

We now define $g(t)$ equal to f except that its zero Fourier components are

$$g(t)_j^{(0)} = c_j\{y^{1/2}(\omega_t y^{-i\sigma}), -y^{3/2}\,\partial_y(\omega_t y^{-i\sigma})\} .$$

Then $g(t)$ is again outgoing for $t \geq 0$ and in addition $g(t)$ is of finite G-norm. Recalling that $Af = i\sigma f$, it is easy to verify that

$$Ag = i\sigma g + \partial_t g$$

and it follows from this that $g \exp(i\sigma t)$ is a solution of the wave equation:

$$\partial_t(e^{i\sigma t}g) = A(e^{i\sigma t}g)$$

and hence that

$$\text{(8.32)} \qquad U(t)\,g(0) = e^{i\sigma t}g(t) \qquad \text{for } t \geq 0 .$$

Since $U(t)$ is unitary with respect to the energy form we may conclude from (8.32) that

$$\text{(8.33)} \qquad E(g(t)) = \text{const.} \qquad \text{for } t \geq 0 .$$

Writing E in the form (8.7)′ and using the Parseval relation to separate out the zero Fourier component in each of the cusp neighborhoods, we see that

$$E(g(t)) = \text{const.} + \sum_{j=1}^{n} \int_a^\infty \left\{ y \left| \partial_y \left[\frac{(g_1)_j^{(0)}}{\sqrt{y}} \right] \right|^2 + \frac{\left|(g_2)_j^{(0)}\right|^2}{y^2} \right\} dy \,.$$

Now ω_t is identically one for $a \le y \le ae^t$ and therefore

$$E(g(t)) \geq \text{const.} + 2\sigma^2 \sum |c_j|^2 \int_a^{ae^t} |y^{-\frac{1}{2}-i\sigma}|^2 \, dy$$

$$\geq \text{const.} + 2\sigma^2 t \sum |c_j|^2$$

which contradicts (8.33) if $\sigma \neq 0$; thus the remaining case of z real, nonzero, and some $c_j \neq 0$ is simply not possible. This completes the proof of Theorem 8.4. ■

COROLLARY 8.5. *Suppose that* f *satisfies the equation*

$$Af = i\sigma f, \qquad \sigma \text{ real} \,,$$

and suppose that $f - f^{(0)}$ *has finite* G-*norm. Then* f *is a linear combination of the generalized eigenfunctions* $e^k(\sigma)$, *constructed above, and of the proper eigenfunctions of* A *with eigenvalues* $i\sigma$.

PROOF. Since f is a solution of the differential equation above, it follows that the zero Fourier component of f in the jth cusp neighborhood is of the form:

$$f_j^{(0)}(y) = (b_j y^{\frac{1}{2}+i\sigma} + c_j y^{\frac{1}{2}-i\sigma})\{1, i\sigma\} \quad \text{for} \quad y > a \,.$$

Making use of (8.27) we conclude that

$$f - \sum \frac{\sqrt{2}\, i\sigma}{a^{-i\sigma}} b_j e^j(\sigma)$$

is outgoing. The conclusion of Corollary 8.5 then follows from Theorem 8.4. ■

We shall study the asymptotic distribution of the eigenvalues of A by means of the calculus of variations. For elliptic operators over a compact domain this method goes back to Herman Weyl (see Courant-Hilbert, [2]); it is somewhat surprising that the same ideas can be used in the present situation where the domain is not compact and where A has both a point and continuous spectrum. What makes it possible is the fact that the contribution of the continuous spectrum to the distribution of the eigenvalues can be accounted for by the winding number of the determinant of the scattering matrix $\mathcal{S}'(\sigma)$.

THEOREM 8.6. *Let* $N(\lambda)$ *denote the number of purely imaginary eigenvalues of* A *of absolute value* $< \lambda$ *and let* $M(\lambda)$ *denote the variation in the winding number of* $\det \mathcal{S}'(\sigma)$ *over the interval* $(-\lambda, \lambda)$. *Then*[1]

$$\lim_{\lambda \to \infty} \frac{N(\lambda) + M(\lambda)}{\lambda^2} = \frac{1}{2\pi} \text{Area } F \,. \tag{8.34}$$

The proof of this theorem is rather lengthy and to make the exposition a bit simpler we shall break it up into four steps.

Step 1. A reformulation in terms of the operator L. As shown in Theorem 3.2 there is a simple relation between the point spectrum of L and that of A: If $-\sigma^2$ *is in the point spectrum of* L *then* $\pm i\sigma$ *are in the point spectrum of* A *and vice versa.* Moreover since L is self-adjoint and has only real coefficients we can, without loss of generality, restrict our attention to real eigenfunctions.

The operator L was defined by means of the form C given in (4.9) as

$$C(u, v) = \iint_F \left\{ u_x v_x + u_y v_y + \frac{uv}{y^2} \right\} dx\, dy \,. \tag{8.35}$$

[1] After having proved this theorem, we learned from R. Gangolli that A. Selberg, in an unpublished manuscript, obtained a similar result from the trace formula.

It is not difficult to prove, using Theorem 4.4, that ϕ is an eigenfunction of L with eigenvalue $-\sigma^2$ if and only if

(8.36) $$C(\phi, u) = \mu(\phi, u)$$

for all smooth u with bounded support, satisfying the boundary conditions (8.3); here

(8.37) $$\mu = \sigma^2 + 5/4 .$$

We note that the zero Fourier coefficients of ϕ in the cusp neighborhoods satisfy the ordinary differential equation

$$y^2 \partial_y^2 \phi_j^{(0)} + \frac{1}{4} \phi_j^{(0)} = -\sigma^2 \phi_j^{(0)}$$

and are therefore of the form

(8.38) $$\phi_j^{(0)} = b_j y^{\frac{1}{2}+i\sigma} + c_j y^{\frac{1}{2}-i\sigma} \qquad \text{for } y > a .$$

Since ϕ lies in $L_2(F)$ it is clear that $\phi_j^{(0)} \equiv 0$ for all $j \neq 0$ and $y > a$.

Step 2. A related quadratic form. Consider the eigenfunctions of the form C relative to the $L_2(F)$ form over the following class of admissible functions:

$\mathcal{A}$ = set of continuous, piecewise smooth functions in $L_2(F)$ satisfying the boundary conditions (8.3) and having vanishing zero Fourier components for $y \geq a$ in each of the cusp neighborhoods.

LEMMA 8.7. *The set of functions*

$$\mathcal{B} = [u \text{ in } \mathcal{A},\ C(u) \leq 1]$$

is contained in a compact subset of $L_2(F)$.

PROOF. It follows by the Rellich criterion that $\mathcal{B}$ is locally L_2 compact. To establish compactness in $L_2(F)$ it therefore suffices to show for each cusp neighborhood that

$$\iint\limits_{y>b} \frac{|u_j|^2}{y^2}\, dx\, dy$$

goes to zero as $b \to \infty$ uniformly in $\mathcal{B}$. Making use of the Parseval relation we see that

$$\iint\limits_{y>b} |u_j|^2\, dx\, dy \leq \sum 4\pi^2 k^2 \int\limits_{y>b} |u_j^{(k)}|^2\, dy = \iint\limits_{y>b} |\partial_x u_j|^2\, dx\, dy .$$

Hence

$$\iint\limits_{y>b} \frac{|u_j|^2}{y^2}\, dx\, dy \leq \frac{1}{b^2} \iint\limits_{y>b} |u_j|^2\, dx\, dy \leq \frac{C(u)}{b^2} .$$

This establishes the uniform convergence to zero and completes the proof of Lemma 8.7. ■

It follows from Lemma 8.7 that the quadratic form C defined on $\mathcal{A}$ has a discrete spectrum with respect to the L_2 form. The nth eigenvalue μ_n can be characterized by the minimax principle:

$$\mu_n = \operatorname*{Min}_{S_n} \operatorname*{Max}_{\phi \in S_n} \frac{C(\phi, \phi)}{(\phi, \phi)} ,$$

where S_n ranges over the n-dimensional subspaces of $\mathcal{A}$.

The eigenfunction ϕ associated with μ satisfies the condition

$$\phi_j^{(0)} = 0 \quad \text{for} \quad y > a \quad \text{and} \quad j = 1, 2, \cdots, n ,$$

and the variational condition

$$C(\phi, u) = \mu(\phi, u)$$

for all u in $\mathcal{A}$. Using elliptic theory we deduce that ϕ satisfies the boundary conditions (8.3) and the partial differential equation

(8.39) $$L\phi = \left(\frac{5}{4} - \mu\right)\phi$$

except along the line $y = a$. The nonzero Fourier components of ϕ are smooth across $y = a$; the zero Fourier components of ϕ vanish for $y \geq a$. Conversely, all functions ϕ of the above kind are eigenfunctions of the problem.

The zero Fourier component of a genuine, that is L_2, eigenfunction of L vanishes in each cusp neighborhood; it therefore follows that each genuine eigenfunction of L is an eigenfunction of our problem. But these are not all; additional eigenfunctions ϕ are obtained from those solutions g of

(8.39)′ $$Lg = \left(\frac{5}{4} - \mu\right) g$$

which have the properties:

i) $g - g^{(0)}$ has finite G norm

and

ii) $g_j^{(0)}(a) = 0$.

ϕ is defined by truncation, that is, ϕ is set equal to g except in the cusp neighborhoods for $y > a$ where

$$\phi_j = g_j - g_j^{(0)} .$$

According to Corollary 8.5, data f of the form $f = \{g, i\sigma g\}$ where g satisfies the differential equation (8.39)′ and property (i) is a linear combination of the generalized eigenfunctions $e^k(\sigma)$ with $\sigma^2 = \mu - 5/4$;

(8.40) $$f = \sum b_k e^k(\sigma) .$$

It follows from (8.27) that

$$f_j^{(0)}(y) = \text{const.} \left(b_j y^{\frac{1}{2}+i\sigma} + \sum_k s_{kj} b_k y^{\frac{1}{2}-i\sigma}\right) \{1, i\sigma\} .$$

Condition (ii) requires that

$$b_j + \left(\sum_k s_{kj}(\sigma)\, b_k\right) a^{-2i\sigma} = 0 \quad \text{for} \quad j = 1, 2, \cdots, n \ .$$

Using (8.30) we can express this condition in the language of matrix theory as follows:

The number 1 *is an eigenvalue of the scattering matrix* $\mathcal{S}'(\sigma)$.

Step 3. Let $P(\mu)$ *denote the number of eigenvalues less than* μ *for the problem described in Step 2. Then*

$$\lim_{\mu \to \infty} \frac{P(\mu)}{\mu} = \frac{1}{4\pi} \text{ Area } F \ . \tag{8.41}$$

PROOF. We obtain upper and lower bounds for $P(\mu)$ by respectively restricting and extending the class of admissible functions. First we consider the restricted class.

$\mathcal{A}_b$ = set of continuous piecewise smooth functions satisfying the boundary conditions (8.3) and, in each cusp neighborhood, vanishing along $y = a$ and for $y > b$ and having vanishing zero Fourier components for $y \geq a$.

Obviously $\mathcal{A}_b \subset \mathcal{A}$ and consequently the corresponding distribution function $P_b(\mu)$ satisfies the relation

$$P_b(\mu) \leq P(\mu) \ . \tag{8.42}$$

Next we define

$\mathcal{A}'_b$ = set of continuous, piecewise smooth functions satisfying the boundary conditions (8.3) and, in each cusp neighborhood, vanishing along $y = a$ and for $y \geq b$.

The eigenfunctions of C over $\mathcal{A}'_b$ are just the eigenfunctions of L on the compact domain $F(a)$, obtained from F by cutting off all n cusp neighborhoods at $y = a$, and the n cylinders $\{a < y < b, |x| < 1/2\}$; Dirichlet boundary conditions are imposed at the ends $y = a$ and $y = b$.

Denote the number of eigenvalues less than μ in this new problem by $P'_b(\mu)$; according to the classical theory of distribution of eigenvalues for second order operators on compact domains

$$(8.41)' \qquad \lim_{\mu\to\infty} \frac{P'_b(\mu)}{\mu} = \frac{1}{4\pi} \text{ Area } F(b) \ ;$$

$F(b)$ is the union of $F(a)$ and the n cylinders.

Next *we claim that*

$$P'_b(\mu) = P_b(\mu) + O(\mu^{1/2}) \ .$$

PROOF. The eigenvalues which occur in $P'_b(\mu)$ include all of those which occur in $P_b(\mu)$ and in addition include all of those corresponding to eigenfunctions in the n cylinders $a < y < b$ whose only nonvanishing Fourier component is the zero component. We obtain an upper bound on the number of such eigenfunctions with values less than μ by replacing the C and L_2 forms for the zero Fourier components in the cusp neighborhoods by lower and upper bounds over (a, b), respectively, for example by

$$C'(u, v) = \sum_j \int_a^b \left(\partial_y u \partial_y v + \frac{uv}{b^2}\right) dy$$

and

$$(u, v)' = \sum_j \int_a^b \frac{uv}{a^2} dy \ .$$

The associated eigenfunctions in each cusp neighborhood are of the form

$$u = \sin \sqrt{\frac{\mu}{a^2} - \frac{1}{b^2}} \, (y-a) \, , \quad (b-a) \sqrt{\frac{\mu}{a^2} - \frac{1}{b^2}} = \pi k \, , \quad k = 1, 2, \cdots \, ,$$

and the number of corresponding eigenvalues less than μ is of $O(\mu^{1/2})$; this proves our claim.

From this estimate and (8.41)′ we deduce

$$\lim_{\mu\to\infty} \frac{P_b(\mu)}{\mu} = \frac{1}{4\pi} \text{ Area } F(b) .$$

Combining this with the inequality (8.42) we see that for all $b > a$,

$$\liminf_{\mu\to\infty} \frac{P(\mu)}{\mu} \geq \frac{1}{4\pi} \text{ Area } F(b) .$$

Since Area $F(b)$ tends to Area F as $b \to \infty$, we deduce that

$$\liminf_{\mu\to\infty} \frac{P(\mu)}{\mu} \geq \frac{1}{4\pi} \text{ Area } F . \tag{8.43}$$

Next we consider an extended class of admissible functions:

$\mathfrak{A}''_b$ = set of piecewise smooth functions satisfying the boundary conditions (8.3), discontinuous along the lines $y = a, b, b+1, b+2, \cdots$ for each cusp neighborhood and for which the zero Fourier components vanish for $y > b$. In this case $\mathfrak{A} \subset \mathfrak{A}''_b$ and hence

$$P(\mu) \leq P''_b(\mu) . \tag{8.44}$$

The previous compactness argument holds for this extended class of admissible functions so that the minimax theory is applicable. We break up the problem into two parts:

$P''_{1b}(\mu)$ = number of eigenvalues $< \mu$ for functions in $\mathfrak{A}''_b$ vanishing for $y > b$ in each cusp neighborhood;

$P''_{2b}(\mu)$ = number of eigenvalues $< \mu$ for functions in $\mathfrak{A}''_b$ vanishing for $y < b$ in each cusp neighborhood.

Then clearly

$$P''_b(\mu) = P''_{1b}(\mu) + P''_{2b}(\mu) . \tag{8.45}$$

Again the asymptotic behavior of $P''_{1b}(\mu)$ is classical:

$$\lim_{\mu \to \infty} \frac{P''_{1b}(\mu)}{\mu} = \frac{1}{4\pi} \text{ Area } F(b) . \tag{8.46}$$

In order to estimate $P''_{2b}(\mu)$ it suffices to consider the eigenfunctions of (8.39) over the square $\{|x| < 1/2,\ c < y < c+1\}$ with periodic conditions on the sides and Neumann conditions top and bottom. We obtain a lower bound on the eigenvalues by replacing y^2 in the C and L_2 forms by $(c+1)^2$ and c^2 respectively. The eigenvalues can then be computed explicitly. Since the zero Fourier component is assumed to be zero we omit the constant eigenfunction. The number of remaining eigenvalues less that μ is majorized by const. $\frac{\mu}{c^2}$. Summing over $c = b, b+1, b+2, \cdots$ we get

$$P''_{2b}(\mu) \le \text{const. } \mu/b .$$

Combining this with (8.45) and (8.46) we obtain

$$\limsup_{\mu \to \infty} \frac{P''_b(\mu)}{\mu} \le \frac{1}{4\pi} \text{ Area } F + \frac{\text{const.}}{b} . \tag{8.47}$$

Combining this with (8.44) we get

$$\limsup_{\mu \to \infty} \frac{P(\mu)}{\mu} \le \frac{1}{4\pi} \text{ Area } F + \frac{\text{const.}}{b} .$$

Since this is true for all b, it follows that

$$\limsup_{\mu \to \infty} \frac{P(\mu)}{\mu} \le \frac{1}{4\pi} \text{ Area } F .$$

This together with (8.43) implies (8.41).

Step 4. We have seen in Step 2 that the eigenfunctions of the quadratic form C over the space $\mathcal{A}$ include in addition to the genuine eigenfunctions of L also the generalized eigenfunctions corresponding to those points σ of the continuous spectrum for which 1 is an eigenvalue of the scattering matrix $\mathcal{S}'(\sigma)$. In this step we show how to relate the number of such σ of absolute value less than λ to $M(\lambda)$, the winding number of $\det \mathcal{S}'(\sigma)$.

REMARK. $M(\lambda)$ was defined as the winding number of $\det \mathcal{S}'(\sigma)$ from $-\lambda$ to λ. According to Proposition 8.11 in the appendix to this section,

$$\mathcal{S}'(\sigma) = \overline{\mathcal{S}'(-\sigma)} \ ;$$

this shows that the winding number of $\det \mathcal{S}'(\sigma)$ over $-\lambda$ to λ is twice the winding number over 0 to λ.

LEMMA 8.8. *Choose* $0 < \kappa < \lambda_j$ *for all relevant* λ_j. *Then in the strip* $-\kappa < \operatorname{Im} z \leq 0$, $\mathcal{S}'(z)$ *is a contraction, that is*

$$\|\mathcal{S}'(z)\| \leq 1 \ , \tag{8.48}$$

provided that the parameter a *appearing in the definition* (8.30) *of the scattering matrix is sufficiently large.*

PROOF. Set $z = \sigma + i\tau$; it follows from definition (8.30) that

$$\|\mathcal{S}'(z, a)\| = \left(\frac{a}{a_0}\right)^{2\tau} \|\mathcal{S}'(z, a_0)\| \ ; \tag{8.49}$$

here a_0 is our initial choice for a. According to (3.43)

$$\mathcal{S}'(z, a_0) = \mathcal{S}_+^{-1}(z)\mathcal{S}''(z)\mathcal{S}_-^{-1}(z) \ . \tag{8.50}$$

The effect of a is simply to translate the origin by $-\log a/a_0$ in the incoming translation representations and by $\log a/a_0$ in the outgoing translation representations. Since S_- and S_+ respectively map incoming into incoming and outgoing into outgoing translation representations and since they commute with translations, it follows that $\mathcal{S}_-$ and $\mathcal{S}_+$ are not effected by a change in a. According to Theorem 3.9 $\mathcal{S}_+$ and $\mathcal{S}_-$

are Blaschke products and because of causality, $\mathcal{S}''(z)$ is a contraction for $\tau \le 0$. So we obtain from (8.50) the norm estimate

(8.51) $$\|\mathcal{S}'(z, a_0)\| \le b(z)$$

where

$$b(z) = \|\mathcal{S}_+^{-1}(z)\| \, \|\mathcal{S}_-^{-1}(z)\| .$$

It follows by the construction of Blaschke products that the following facts are evident:

1) $b(\sigma) = 1$ for σ real,
2) $b(z) \to 1$ as $|z| \to \infty$,
3) $\partial b/\partial \tau$ is bounded for real σ.

Since by Theorem 3.9, $\mathcal{S}_+^{-1}$ and $\mathcal{S}_-^{-1}$ are holomorphic in the strip $-\kappa \le \operatorname{Im} z \le 0$, we easily deduce from these facts, that for a sufficiently large

$$\left(\frac{a}{a_0}\right)^{2\tau} \le b(\sigma + i\tau), \text{ for all } z \text{ in this strip} .$$

Combining this with (8.49) and (8.51) we verify (8.48). ■

Next we denote the eigenvalues of $\mathcal{S}'(z)$ by $\gamma(z)$ and introduce the notation

$$\log \gamma(z) = \alpha(\sigma,\tau) + i\beta(\sigma,\tau) ,$$

where as before $z = \sigma + i\tau$. Since $\mathcal{S}'(z)$ is unitary for z real, it follows that $|\gamma(\sigma)| = 1$; this implies that

$$\alpha(\sigma, 0) = 0 .$$

Since by Lemma 8.8, $\mathcal{S}'(z)$ is a contraction in the strip $-\kappa \le \tau \le 0$, it follows that $\|\mathcal{S}'(z)\| \le 1$ and so

$$\alpha(\sigma,\tau) \le 0 \quad \text{in this strip} .$$

Combining this with $\alpha(\sigma, 0) = 0$ we deduce

(8.52) $$\partial_\tau \alpha(\sigma,\tau)|_{\tau=0} \ge 0 .$$

LEMMA 8.9. *Each* $\beta(\sigma, 0)$ *is a decreasing function of* σ, *provided that* a *is large enough.*

PROOF. It is clear that $\log \gamma(z)$ is an n-valued analytic function of z and since each $\alpha(z)$ is zero on the real axis, the real axis is free of branch points. By the Cauchy-Riemann equations

$$\partial_\sigma \beta(\sigma, 0) = -\partial_\tau \alpha(\sigma, \tau)|_{\tau=0} \; .$$

Combining this with (8.52) we see that

$$\partial_\sigma \beta(\sigma, 0) \leq 0 \; ,$$

as asserted in Lemma 8.9. ∎

Having shown that $\beta(\sigma, 0)$ changes monotonically, it follows that as σ goes from 0 to λ, each eigenvalue

$$\gamma(\sigma) = e^{i\beta(\sigma)},$$

moves clockwise around the unit circle. The product of the n eigenvalues is $\det \mathcal{S}'(\sigma)$ and therefore the number of times the eigenvalues pass through the value $\gamma = 1$ differs at most by n from the change in the argument of $\det \mathcal{S}'(\sigma)$ as σ goes from 0 to λ; that change in the argument of $\det \mathcal{S}'(\sigma)$ is $\frac{1}{2} M(\lambda)$.

Denote by $Q(\mu)$ the number of eigenvalues of C over $\mathfrak{A}$ which are less than μ and which come from generalized eigenfunctions corresponding to points of the continuous spectrum. To such a point σ there corresponds by (8.37) the eigenvalue $\sigma^2 + 5/4$. The above result can then be stated as follows:

$$|Q(\lambda^2 + 5/4) - \tfrac{1}{2} M(\lambda)| \leq n \; . \tag{8.53}$$

Theorem 8.6 is now an easy consequence of these four steps. By Steps 1 and 2

$$N(\lambda) = 2[P(\lambda^2 + 5/4) - Q(\lambda^2 + 5/4)] \; .$$

By Step 3, see (8.41),

$$\lim_{\lambda \to \infty} \frac{P(\lambda^2 + 5/4)}{\lambda^2} = \frac{1}{4\pi} \text{ Area } F$$

and by Step 4

$$\lim_{\lambda \to \infty} \left[\frac{Q(\lambda^2 + 5/4)}{\lambda^2} - \frac{M(\lambda)}{2\lambda^2} \right] = 0$$

holds for a sufficiently large. Combining these results we obtain (8.34) for a sufficiently large. However by (8.30) it is clear that

$$M(\lambda, a) = M(\lambda, a_0) - 2\lambda n \log a/a_0 \ .$$

Since a linear dependence can not effect (8.34), we conclude that this relation holds for any suitable a. ■

Appendix to Section 8: Functional equations for the Eisenstein series and the scattering matrix.

There are two basic symmetries of the automorphic wave equation which can be used to derive functional equations for the Eisenstein series and the scattering matrix, namely *reality* and *time reversibility*. A third relation for the scattering matrix can be obtained from the fact that the system is *energy conserving*.

As before let $e^k(z)$ denote the generalized eigenfunction of A:

$$Ae^k = iz\, e^k \tag{8.54}$$

with zero Fourier component in the jth cusp neighborhood of the form

$$[e^k(z)]_j^{(0)} = \frac{a^{-iz}}{\sqrt{2}\, iz} (\delta_{kj}\, y^{\frac{1}{2}+iz} + s_{kj}(z)\, y^{\frac{1}{2}-iz})\{1, iz\} \ . \tag{8.27$'$}$$

According to Theorem 8.3 the scattering matrix $\mathcal{S}'(z)$ is

$$\mathcal{S}'_{jk}(z) = -a^{-2iz} s_{kj}(z) \ . \tag{8.30$'$}$$

PROPOSITION 8.10. *Reality implies that*

$$e^k(z) = \overline{e^k(-\bar{z})} \tag{8.55$_1$}$$

and

$$\mathcal{S}'(z) = \overline{\mathcal{S}'(-\bar{z})} \; . \tag{8.55$_2$}$$

PROOF. From the fact that A is a real operator we get by conjugating (8.54):

$$A\overline{e^k(z)} = -i\bar{z}\,\overline{e^k(z)} \; ;$$

and from (8.27)′ we see that the zero Fourier component of $\overline{e^k(-\bar{z})}$ in the jth cusp neighborhood is given by

$$[\overline{e^k(-\bar{z})}]_j^{(0)} = \frac{a^{-iz}}{\sqrt{2}\,iz}\,(\delta_{kj}\,y^{\frac{1}{2}+iz} + \overline{s_{kj}(-\bar{z})}\,y^{\frac{1}{2}-iz})\{1, iz\} \; . \tag{8.56}$$

As we have seen in Lemma 7.10 and Theorem 7.11, only the zero Fourier component of $e^k(z)$ has infinite energy; in particular

$$e(z) = e^k(z) - \overline{e^k(-\bar{z})} \tag{8.57}$$

is an eigenfunction of A with vanishing $\mathcal{D}'_-$ component and hence of finite energy for $\operatorname{Im} z < 0$. Lemma 7.13 therefore implies that $e(z) = 0$ if $\operatorname{Im} z < 0$ and iz is not in the point spectrum of A; by continuity even these exceptional values can be ignored. Since both terms on the right in (8.57) are locally analytic in z we conclude that $(8.55)_1$ holds for all z which are not poles of $\mathcal{S}'(z)$. Moreover if we compare (8.27)′ and (8.56) we obtain

$$s_{kj}(z) = \overline{s_{kj}(-\bar{z})}$$

which together with (8.30) implies $(8.55)_2$. ∎

Denoting the time reversal operator by R as in (3.39), we have

PROPOSITION 8.11. *Time reversibility implies that*

$$(8.58)_1 \qquad e^k(z) = \sum_{j=1}^{n} R\, e^j(-z)\,[\mathcal{S}'(-z)]^{-1}_{jk}$$

and

$$(8.58)_2 \qquad \mathcal{S}'(z) = [\mathcal{S}'(-z)]^{-1} .$$

PROOF. Setting

$$(t_{kj}(z)) = (s_{kj}(z))^{-1} ,$$

we see by (8.30) that

$$(8.59) \qquad t_{kj}(z) = -a^{-2iz}\,[\mathcal{S}'(z)]^{-1}_{jk} .$$

Since R anticommutes with $A : RA = -AR$, it follows that

$$AR\,e^k(z) = -iz\,R\,e^k(z)$$

and hence $Re^k(z)$ is a generalized eigenfunction of A with corresponding eigenvalue $-iz$. Multiplying (8.27)′, with $-z$ in place of z, by $t_{\ell k}(-z)$ and summing over k we get

$$\sum_k t_{\ell k}(-z)\,[e^k(-z)]^{(0)}_j = \frac{-a^{iz}}{\sqrt{2}\,iz}\,(\delta_{\ell j}\, y^{\frac{1}{2}+iz} + t_{\ell j}(-z)\, y^{\frac{1}{2}-iz})\{1, -iz\} .$$

Proceeding as in the proof of Proposition 8.10 we see that

$$e(z) = e^{\ell}(z) + a^{-2iz} \sum_k t_{\ell k}(-z)\, Re^k(-z)$$

is a generalized eigenfunction of A which is of finite energy for $\operatorname{Im} z < 0$. Again by Lemma 7.13, $e(z)$ vanishes for $\operatorname{Im} z < 0$ and so does its analytic extension into the complex plane minus the poles of $\mathcal{S}'(z)$. Using (8.59) to substitute for $t_{\ell k}(-z)$ we obtain $(8.58)_1$ and comparing the zero Fourier component coefficients of $y^{1/2-iz}$ in each cusp neighborhood gives $(8.58)_2$. ∎

PROPOSITION 8.12. *Conservation of energy implies that*

$$[\mathcal{S}'(\bar{z})]^* = [\mathcal{S}'(z)]^{-1} . \tag{8.60}$$

PROOF. This is an immediate consequence of Theorem 2.6, part (ii), according to which $\mathcal{S}'(z)$ is unitary for real z, that is

$$[\mathcal{S}'(\bar{z})]^* \mathcal{S}'(z) = I \qquad \text{for real } z .$$

Analytic continuation gives the desired result. ■

COROLLARY 8.13. Reciprocity. *The matrix* $\mathcal{S}'$ *is symmetric*:

$$\mathcal{S}'_{jk}(z) = \mathcal{S}'_{kj}(z) . \tag{8.61}$$

PROOF. Combining the relations $(8.55)_2$, $(8.58)_2$ and (8.60) gives

$$\mathcal{S}'_{jk}(z) = \overline{\mathcal{S}'_{jk}(-\bar{z})} = \overline{[\mathcal{S}'(\bar{z})]^{-1}_{jk}} = \mathcal{S}'_{kj}(z)$$

as desired. ■

In case of the modular group, the relation $(8.58)_1$ can be used to obtain another proof of the well-known functional equation for the Riemann zeta function; see Kubota [p. 46, 12].

PROPOSITION 8.14. $\mathcal{S}'(0) = I$.

PROOF. Since the analogue of Theorem 7.11 holds, we know that the generalized eigenfunction $e^k(z)$ is meromorphic in the local sense with poles contained in $-i\sigma(B'')$ and the relevant $\pm i\lambda_j$. In particular $e^k(z)$ is holomorphic at $z = 0$. It is clear from $(8.27)'$ that this requires $s_{jk}(0) = -\delta_{jk}$ so that $\mathcal{S}'(0) = I$. ■

§9. THE SELBERG TRACE FORMULA

The trace formula, discovered by Selberg [21] in 1956, has played an important role in the theory of automorphic functions. In this section we shall present two versions of this formula based on the wave equation. In the body of the section we use the wave operator U to derive the original Selberg version; we believe this proof is more direct than previous proofs although the general pattern of the original Selberg argument remains. The appendix contains a new version of the identity based on the associated semigroup of nonself-adjoint operators Z. In this version a sum over the eigenvalues of the outgoing generalized eigenfunctions of A replaces an integral in the original trace formula.

A self-adjoint operator K on a Hilbert space is said to be of *trace class* if it has a discrete spectrum and if the eigenvalues $\{\kappa_j\}$ of K satisfy the condition

$$\sum |\kappa_j| < \infty \,. \tag{9.1}$$

It is easy to show that if we represent an operator of trace class as an infinite matrix with respect to any orthonormal basis $\{f_j\}$, then the trace of this matrix is independent of the choice of the orthonormal basis; this quantity is called the *trace* of K:

$$\operatorname{tr} K = \sum (Kf_j, f_j) \,. \tag{9.2}$$

In particular if we choose the f_j to be the normalized eigenvectors: $Kf_j = \kappa_j f_j$, then $(Kf_j, f_j) = \kappa_j$ and we see that the trace can be represented as the sum of the eigenvalues of K:

$$\operatorname{tr} K = \sum \kappa_j \,. \tag{9.3}$$

The following facts about trace class operators and the trace are well known and not hard to prove (see, for instance, Gohberg and Krein [11]):

i) A bounded operator K is of trace class if the sum

$$\sum (Kf_j, f_j)$$

converges for every orthonormal basis.

ii) The trace class self-adjoint operators form a real linear space on which the trace is a linear functional.

iii) If K is of trace class and T is a bounded operator, then TK and KT are also of trace class and $\operatorname{tr}(TK) = \operatorname{tr}(KT)$.

iv) If $\{T_n\}$ is a sequence of operators tending to the identity in the strong topology, then

$$\lim \operatorname{tr}(T_n K T_n) = \operatorname{tr} K \quad (9.4)$$

v) Suppose that K is an integral operator of trace class acting on L_2 functions defined on some domain F:

$$(Kf)(w) = \int_F K(w,w')\, f(w')\, dw'$$

with kernel $K(w,w')$ which is continuous over $F \times F$. In this case the trace of K can be evaluated by integrating the kernel along the diagonal:

$$\operatorname{tr} K = \int_F K(w,w)\, dw \quad (9.5)$$

We return now to the automorphic wave equation

$$u_{tt} = y^2 \Delta u + \frac{1}{4} u \ ;$$

and, as before, we denote by $U(t)$ the solution operator relating initial data of solutions automorphic with respect to some discontinuous subgroup Γ of G to their data at time t. $U(t)$ and $U(-t)$ are adjoint to each other relative to the E-form; consequently $U(t)+U(-t)$ is a self-adjoint operator in the E-form. However it is not of trace class for two reasons:

a) its continuous spectrum is nonempty;

b) its point spectrum does not satisfy the condition (9.1).

The remedy for the first defect is simply the removal of the continuous spectrum; that is, we consider

$$U_p(t) + U_p(-t) \tag{9.6}$$

where U_p is U projected onto the subspace of $\mathcal{H}$ spanned by the proper eigenvectors of U. Note that

$$U_p = U - U_c \tag{9.7}$$

where U_c is the projection of U onto the subspace $\mathcal{H}'_c$ spanned by the generalized eigenvectors of U.

To make a trace class operator out of (9.6) we multiply this operator by a smooth rapidly decreasing function $\phi(t)$ and integrate; more precisely, we consider the operator

$$C_p = \int_{-\infty}^{\infty} \phi(t)\, U_p(t)\, dt \tag{9.8}$$

where $\phi(t)$ is a *real valued, smooth, rapidly decreasing, even function of* t.

We must also face up to the fact that the E-form is not positive definite on the underlying Hilbert space $\mathcal{H}$; to take care of this we make use of the *time reversal* operator R defined in Section 3 as

$$R\{f_1, f_2\} = \{f_1, -f_2\} . \tag{9.9}$$

We recall that R intertwines U(t) and U(−t):

$$R\,U(t) = U(-t)\,R\ .$$

It follows easily from this that

$$R\,U_p(t) = U_p(-t)\,R\ ,$$

from which we deduce that R *commutes with* $U_p(t) + U_p(-t)$. But then so does the operator Q defined as

$$Q = \frac{1}{2}(I - R)\ ;$$

Q is the orthogonal projection onto the second component of the data:

(9.10) $$Q\{f_1, f_2\} = \{0, f_2\}\ .$$

Since Q commutes with $U_p(t) + U_p(-t)$, it follows that the range of Q is an invariant subspace of this operator. Likewise for every even function ϕ the range of Q is an invariant subspace for the operator C_p defined by (9.8).

Now the range of Q consists of all data of the form $\{0, f_2\}$; we denote this subspace by $\mathcal{H}_2$. It is obvious that the energy form is positive definite on $\mathcal{H}_2$ where it becomes the L_2 norm of f_2 over F. *In what follows we shall study the trace of the operator* C_p *defined by* (9.8) *and restricted to* $\mathcal{H}_2$.

The spectrum of C_p restricted to $\mathcal{H}_2$ is easily related to the point spectrum of the Laplace-Beltrami operator

$$L = y^2\Delta + \frac{1}{4}\ .$$

We have seen in Theorem 3.2 that if $\kappa \neq 0$ is a simple eigenvalue of L then $\pm\sqrt{\kappa}$ are eigenvalues of A and $\exp(\pm\sqrt{\kappa}\,t)$ are eigenvalues of U(t). Moreover $\exp(\sqrt{\kappa}\,t) + \exp(-\sqrt{\kappa}\,t)$ is an eigenvalue of $U(t) + U(-t)$ of multiplicity 2 over $\mathcal{H}$ and of multiplicity 1 over $\mathcal{H}_2$. If $\kappa = 0$ is an eigenvalue of L paired with the eigenfunction q, then since $E(\{q,0\}) = 0$ we see that $\{0,q\}$ is annihilated by A and hence that 1 is an eigenvalue of U(t). In either case the eigenvalue of the operator C_p corresponding to κ is

$$\int_{-\infty}^{\infty} e^{\sqrt{\kappa}t}\phi(t)\,dt \;. \tag{9.11}$$

We have seen in Corollary 6.14 that the operator L has a finite number of positive eigenvalues λ_j^2, all $\leq 1/4$; a null space of dimension $\eta < \infty$; and an infinite sequence of negative eigenvalues $-\nu_j^2$, $\nu_j > 0$, each of finite multiplicity.

Denote by Φ the Fourier transform of ϕ

$$\Phi(\sigma) = \int_{-\infty}^{\infty} e^{i\sigma t}\phi(t)\,dt \;; \tag{9.12}$$

Φ will be a real valued even function. It follows from (9.11) that the spectrum of the operator C_p consists of the numbers

$$\Phi(i\lambda_j),\ \Phi(0)\ \ \eta\ \text{times},\ \ \Phi(\nu_j)\ . \tag{9.13}$$

In order for $\Phi(i\lambda_j)$ to make sense we must impose on ϕ a condition of exponential decay. Since none of the λ_j^2 exceed 1/4, a decay of the form

$$|\phi(t)| \leq \text{const.}\ e^{-(\frac{1}{2}+\delta)|t|}, \quad \delta > 0\ , \tag{9.14}$$

suffices.

The sum of the absolute values of the eigenvalues (9.13) is

$$\sum |\Phi(i\lambda_j)| + \eta|\Phi(0)| + \sum |\Phi(\nu_j)|\ . \tag{9.15}$$

To assure the convergence of the infinite series standing in the third place in (9.15) we have to impose a further decay condition on Φ. According to Theorem 8.7 the number $N(\sigma)$ of ν_j not exceeding σ is of order σ^2; that is

$$N(\sigma) = O(\sigma^2) . \tag{9.16}$$

We write the difference of the partial sums of the series in (9.15) as an integral:

$$\sum_{a<\nu_j<\beta} |\Phi(\nu_j)| = \int_a^\beta |\Phi(\tau)|\, dN(\tau) .$$

Integration by parts gives

$$\sum_{a<\nu_j<\beta} |\Phi(\nu_j)| = |\Phi(\tau)| N(\tau) \Big|_a^\beta - \int_a^\beta \frac{d|\Phi(\tau)|}{d\tau} N(\tau)\, d\tau . \tag{9.17}$$

It follows from (9.16) that the expression on the right in (9.17) tends to zero as $a, \beta \to \infty$ if Φ satisfies

$$\left|\frac{d\Phi(\tau)}{d\tau}\right| \leq \frac{\text{const.}}{(1+|\tau|)^{3+\delta}} , \quad \delta > 0 . \tag{9.18}$$

Thus condition (9.18) guarantees that the operator C_p is of trace class; its trace is

$$\operatorname{tr} C_p = \sum \Phi(i\lambda_j) + \eta \Phi(0) + \sum \Phi(\nu_j) . \tag{9.19}$$

Note that (9.18) implies that ϕ must be C^1.

Next we evaluate the trace of C_p in another way, employing the following limiting process: We define the operator T_Y as *truncation* in each cusp above the level Y; that is, for any function f defined on the fundamental domain F we define $T_Y f$ by

$$T_Y f = \chi_{F_0(Y)} f ,$$

where $\chi_{F_0(Y)}$ is the characteristic function of the set $F_0(Y)$ which is defined in the terminology of (8.1) as

$$F_0(Y) = F \setminus \bigcup_j g_j[F_1 \quad \{y > Y\}] .$$

The operators T_Y are orthogonal projections on $L_2(F)$ which tend strongly to the identity as $Y \to \infty$. As observed at the beginning of this section, the operators $T_Y C_p T_Y$ are also of trace class and

(9.20) $$\lim_{Y \to \infty} \operatorname{tr}(T_Y C_p T_Y) = \operatorname{tr} C_p .$$

Substituting the decomposition (9.7) of U into the definition (9.8) of C_p we get the following decomposition of C_p:

$$C_p = C - C_c$$

where

(9.21) $$C = \int \phi(t)\, U(t)\, dt$$

and

$(9.21)_c$ $$C_c = \int \phi(t)\, U_c(t)\, dt \ ;$$

as before all of these operators are restricted to $\mathcal{H}_2$. From this decomposition we deduce that

(9.22) $$T_Y C_p T_Y = T_Y C\, T_Y - T_Y C_c T_Y .$$

We shall subsequently show that the operator $T_Y C_c T_Y$ is of trace class. It therefore follows by property (ii), stated at the beginning of this section, that $T_Y C\, T_Y$ is also of trace class and that

(9.23) $$\operatorname{tr}(T_Y C_p T_Y) = \operatorname{tr}(T_Y C\, T_Y) - \operatorname{tr}(T_Y C_c T_Y) .$$

We shall evaluate the traces on the right in (9.23) separately. For the second term this is relatively easy since we can write C_c fairly explicitly as an integral operator by using the spectral representation of the absolutely continuous part A_c of the generator A, derived in Theorem 8.1.

We can write

$$U_c(t) = e^{A_c t} .$$

Multiplying this by $\phi(t)$, integrating and using the definition (9.12) of Φ, we obtain

$$C_c = \int \phi(t)\, U_c(t)\, dt = \int \phi(t)\, e^{A_c t}\, dt \tag{9.24}$$

$$= \int \phi(t)\, e^{i(-iA_c)t}\, dt = \Phi(-iA_c) ,$$

restricted to $\mathcal{H}_2$. According to Theorem 8.1 the operator $-iA_c$ can be represented as an integral operator with kernel

$$\ker(-iA_c) = \sum_j \frac{1}{2\pi} \int \sigma\, e_j(\sigma, w)\, e_j(\sigma, w')\, d\sigma ,$$

where the $e_j(\sigma, w)$ are the generalized eigenfunctions described in Section 8. There are n of them, n being the number of inequivalent cusps as well as the multiplicity of the continuous spectrum. For any function of $-iA_c$ we have the corresponding formula for its kernel:

$$\ker(\Phi(-iA_c)) = \sum_j \frac{1}{2\pi} \int \Phi(\sigma)\, e_j(\sigma, w)\, e_j(\sigma, w')\, d\sigma . \tag{9.25}$$

The operator C_c is the restriction of $\Phi(-iA_c)$ to data whose first component is zero; this can be accomplished by replacing e_j in the above formula by the second component only of e_j. Next we apply T_Y

to the right and to the left of C_c. Using the definition of T_Y and (9.25) we get

$$\ker(T_Y C_c T_Y) = \sum_j \frac{1}{2\pi} \int \Phi(\sigma)\, e_{j,Y}(\sigma, w)\, e_{j,Y}(\sigma, w')\, d\sigma , \tag{9.26}$$

where

$$e_{j,Y}(\sigma, w) = \chi_{F_0(Y)}(w)\, e_{j2}(\sigma, w) ; \tag{9.27}$$

here e_{j2} denotes the second component of e_j.

We shall prove that $T_Y C_c T_Y$ is of trace class and at the same time evaluate its trace by using formula (9.2). Let f be any function in $L_2(F)$; denoting the $L_2(F)$ scalar product by parentheses we see from (9.26) that

$$(T_Y C_c T_Y f, f) = \sum_j \frac{1}{2\pi} \int \Phi(\sigma)\, |(e_{j,Y}(\sigma), f)|^2\, d\sigma .$$

Summing over any orthonormal basis $\{f_k\}$ we may conclude by property (i), noted at the beginning of this section, and by (9.2) that $T_Y C_c T_Y$ is of trace class with trace

$$\operatorname{tr}(T_Y C_c T_Y) = \sum_j \frac{1}{2\pi} \int \Phi(\sigma)\, \|e_{j,Y}(\sigma)\|^2\, d\sigma , \tag{9.28}$$

provided that the integrals on the right converge absolutely; here $\|e\|$ denotes the $L_2(F)$ norm of e. The absolute convergence of these integrals is a simple consequence of the following evaluation of them since we can replace Φ by its absolute value everywhere in the argument.

We begin by evaluating the right side of (9.28) approximately for large Y. To this end we introduce the quantity e_j^Y, obtained by truncating above Y in each cusp neighborhood only the zero Fourier component of e_{j2}:

$$e_j^Y(\sigma, w) = \begin{cases} e_{j2}(\sigma, w) & \text{in} \quad F_0(Y) \\ e_{j2}(\sigma, w) - e_{j2}^{(0)}(\sigma, w) & \text{outside} \quad F_0(Y) . \end{cases} \tag{9.29}$$

Although e_j^Y does not differ very much from $e_{j,Y}$, its norm is easier to evaluate; the facts are contained in the following two lemmas:

LEMMA 9.1.

a) *For all* σ *and all* Y

(9.30) $$0 \le \sum_j (\|e_j^Y(\sigma)\|^2 - \|e_{j,Y}(Y)\|^2) \le \text{const.} + |\partial_\sigma M(\sigma)| ,$$

where $M(\sigma)$ *is the winding number of* $\det \mathcal{S}'(\sigma)$ *introduced in Theorem 8.7.*

b) *For every* j *and* σ

(9.31) $$\lim_{Y\to\infty} \|e_j^Y(\sigma)\|^2 - \|e_{j,Y}(\sigma)\|^2 = 0 .$$

LEMMA 9.2.

(9.32) $$\|e_j^Y(\sigma)\|^2 = \log Y + \frac{i}{2} \sum_k \overline{\mathcal{S}}'_{jk}(\sigma)\partial_\sigma \mathcal{S}'_{jk}(\sigma) - \frac{1}{2\sigma} \operatorname{Im}(\overline{\mathcal{S}}'_{jj}(\sigma)\, Y^{2i\sigma}) .$$

Before proving these lemmas we show how to deduce from them the following:

LEMMA 9.3. *The operator* $T_Y C_c T_Y$ *is of trace class with trace:*

(9.33) $$\operatorname{tr}(T_Y C_c T_Y) = n\phi(0)\log Y$$

$$+ \frac{i}{4\pi} \int \Phi(\sigma)\operatorname{tr}(\mathcal{S}'^*(\sigma)\partial_\sigma \mathcal{S}'(\sigma))\, d\sigma - \frac{1}{4}\Phi(0)\operatorname{tr}\mathcal{S}'(0) + o(1).$$

PROOF. We first show that replacing $\|e_{j,Y}(\sigma)\|^2$ by $\|e_j^Y(\sigma)\|^2$ in (9.28) results in an error which tends to zero as $Y \to \infty$. To see this we make use of properties of $M(\sigma)$ established in Theorem 8.7 where it was proved that $M(\sigma)$ is a monotonic function of σ, if we choose a sufficiently large, and that

(9.34) $$M(\sigma) \leq \text{const.}\ \sigma^2 \ .$$

Therefore we can integrate (9.30) and obtain, after summing with respect to j,

$$\int_0^\sigma \|e^Y(\sigma)\|^2 - \|e_Y(\sigma)\|^2 \, d\sigma \leq \text{const.}\ |\sigma| + \text{const.}\ \sigma^2 \ .$$

This together with the assumed bound (9.18) for $d\Phi/d\sigma$ permits us to write, after an integration by parts,

(9.35) $$\int \Phi(\sigma)(\|e^Y(\sigma)\|^2 - \|e_Y(\sigma)\|^2)\, d\sigma$$

$$= \int \left(\frac{d}{d\sigma}\Phi(\sigma)\right)\left(\int_0^\sigma (\|e^Y(\sigma)\|^2 - \|e_Y(\sigma)\|^2)\, d\tau\right) d\sigma \ ,$$

and shows that the integrand on the right is dominated by

$$\frac{\text{const.}}{(1+|\sigma|)^{1+\delta}} \ .$$

On the other hand, according to (9.31) the integrand in (9.35) tends to zero for each σ as $Y \to \infty$; therefore, by the Lebesgue dominated convergence theorem, the expression (9.35) tends to zero. This proves that the error caused by replacing $\|e_{j,Y}(\sigma)\|^2$ on the right side of (9.28) by $\|e_j^Y(\sigma)\|^2$ tends to zero as $Y \to \infty$.

Making this replacement and using (9.32) we obtain

(9.36) $$\operatorname{tr}(T_Y C_c T_Y) = \sum_j \frac{1}{2\pi} \int \Phi(\sigma)\Big[\log Y$$
$$+ \frac{i}{2} \sum_k \overline{\mathcal{S}}'_{jk}(\sigma)\,\partial_\sigma \mathcal{S}'_{jk}(\sigma) - \frac{1}{2\sigma} \operatorname{Im}(\overline{\mathcal{S}}'_{jj}(\sigma)\, Y^{2i\sigma})\Big] d\sigma + o(1) \, .$$

The first term on the right in (9.36) equals

$$\frac{n}{2\pi} \int \Phi(\sigma)\, d\sigma \log Y = n\phi(0) \log Y .$$

The second term is the integral of $\Phi(\sigma)$ times

$$\frac{i}{4\pi} \operatorname{tr}(\mathcal{S}'^{*}(\sigma)\partial_\sigma \mathcal{S}'(\sigma)) ,$$

a quantity equal to the rate of increase of the winding number $M(\sigma)$ of $\det \mathcal{S}'(\sigma)$. Again making use of (9.18) and (9.34), an integration by parts shows that

$$\frac{i}{4\pi} \int \Phi(\sigma) \operatorname{tr}(\mathcal{S}'^{*}(\sigma)\partial_\sigma \mathcal{S}'(\sigma)\, d\sigma$$

is absolutely integrable.

As for the last integrated term on the right in (9.36), we recall the relation $(8.55)_2$:

$$\mathcal{S}'_{jj}(-\sigma) = \overline{\mathcal{S}}'_{jj}(\sigma) . \tag{9.37}$$

In particular $\mathcal{S}'_{jj}(0)$ is real valued and this together with the regularity of $\mathcal{S}'(z)$ near $z = 0$ implies that $\operatorname{Im} \mathcal{S}'_{jj}(\sigma)/\sigma$ is regular at $\sigma = 0$. Consequently $\operatorname{Im} \overline{\mathcal{S}}'_{jj}(\sigma)\, Y^{2i\sigma}/\sigma$ is also regular at $\sigma = 0$ and since $|\mathcal{S}_{jj}(\sigma)| \leq 1$, it follows that the integral

$$\frac{1}{4\pi} \int \Phi(\sigma) \frac{\operatorname{Im}(\overline{\mathcal{S}}'_{jj}(\sigma)\, Y^{2i\sigma})}{\sigma}\, d\sigma$$

is absolutely convergent. It can be rewritten as

$$\frac{1}{4\pi}\int \Phi(\sigma)\mathrm{Re}\,\mathcal{S}'_{jj}(\sigma)\,\frac{\sin(2\sigma \log Y)}{\sigma}\,d\sigma - \frac{1}{4\pi}\int \Phi(\sigma)\,\frac{\mathrm{Im}\,\mathcal{S}'_{jj}(\sigma)}{\sigma}\cos(2\sigma \log Y)\,d\sigma\ .$$

Both integrals converge to limits as $Y \to \infty$:

$$\lim \frac{1}{4\pi}\int \Phi(\sigma)\mathrm{Re}\,\mathcal{S}'_{jj}(\sigma)\,\frac{\sin(2\sigma \log Y)}{\sigma}\,d\sigma = \frac{1}{4}\,\Phi(0)\mathcal{S}'_{jj}(0)$$

by a known theorem on Fourier integrals; since $\mathrm{Im}\,\mathcal{S}'_{jj}(\sigma)/\sigma$ is regular at $\sigma = 0$,

$$\lim \frac{1}{4\pi}\int \Phi(\sigma)\,\frac{\mathrm{Im}\,\mathcal{S}'_{jj}(\sigma)}{\sigma}\cos(2\sigma \log Y)\,d\sigma = 0$$

by the Riemann-Lebesgue theorem. Combining these results expressing the various terms on the right in (9.36), we obtain (9.33). As noted earlier, the absolute convergence of each of these terms is all that is needed to prove that $T_Y C_c T_Y$ is of trace class. This completes the proof of Lemma 9.3. ■

We turn now to the

PROOF OF LEMMA 9.2. It was shown in Section 8 that after the removal of the zero Fourier coefficients of the generalized eigenfunctions in each cusp neighborhood, the remainder, $e_j^Y(z)$ is a continuous function of z in the energy norm for z near the real axis. Therefore

$$\|e_j^Y(\sigma)\| = \lim_{\tau \to 0} \|e_j^Y(\sigma + i\tau)\|\ .$$

Next we decompose F as

$$F = \bigcup_{k=0}^{n} F_k(Y)$$

where in the terminology of (8.1)

$$F_k(Y) = g_k[F_1 \cap \{y > Y\}] \quad \text{for} \quad k = 1, 2, \cdots, n \ ,$$

and, as before, $F_0(Y)$ is the remainder. Since an integral over F is the sum of the integrals over the $F_k(Y)$, we have

$$\|e_j^Y(z)\|^2 = \sum_{k=0}^{n} \|e_j^Y(z)\|^2_{F_k(Y)} \ . \tag{9.38}$$

In each domain $F_k(Y)$, $e_j^Y(z)$ satisfies the eigenvalue equation

$$L e_j^Y = -z^2 e_j^Y \ . \tag{9.39}$$

Indeed this is obvious in $F_0(Y)$ since in this sub-domain $e_j^Y = e_{j2}$; it is also true in each cusp neighborhood where the equation (9.39) is satisfied by each Fourier component of e_{j2}. We deduce from (9.39) that

$$(\bar{z}^2 - z^2)\|e_j^Y(z)\|^2_{F_k(Y)} = (L e_j^Y(z), e_j^Y(z))_{F_k(Y)} - (e_j^Y(z), L e_j^Y(z))_{F_k(Y)} \ . \tag{9.40}$$

In each subset $F_k(Y)$ we apply Green's formula to the expression on the right in (9.40). Since all but the zeroth component of e_j^Y is continuous across the cuts separating the cusps from $F_0(Y)$, the resulting boundary terms can be written in the form:

$$\sum_{k=1}^{n} \int \left[(\partial_y e_{jk}^{(0)}(z)) \overline{e_{jk}^{(0)}}(z) - e_{jk}^{(0)}(z) \partial_y \overline{e_{jk}^{(0)}}(z) \right]_{y=Y} dx \ . \tag{9.41}$$

According to formula (8.27), the second component of $e_j^{(0)}(z)$ is

$$e_{jk}^{(0)}(z) = \frac{a^{-iz}}{\sqrt{2}} (\delta_{jk} y^{\frac{1}{2}+iz} - S'_{jk}(z) y^{\frac{1}{2}-iz}) \ .$$

Substituting this into (9.40) and (9.41) gives the following:

$$-4\sigma\tau i \|e_j^Y(z)\|^2 = \frac{a^{2\tau}}{2} \sum_{k=1}^{n} \Big\{ \Big[\Big(\tfrac{1}{2} + iz\Big)\delta_{jk} Y^{-\frac{1}{2}+iz} - \Big(\tfrac{1}{2} - iz\Big) \mathcal{S}'_{jk}(z) Y^{-\frac{1}{2}-iz}\Big]$$

$$\times [\delta_{jk} Y^{\frac{1}{2}+iz} - \mathcal{S}'_{jk}(z) Y^{\frac{1}{2}-iz}]^- - [\delta_{jk} Y^{\frac{1}{2}+iz} - \mathcal{S}'_{jk}(z) Y^{\frac{1}{2}-iz}]$$

$$\times \Big[\Big(\tfrac{1}{2} + iz\Big) \delta_{jk} Y^{-\frac{1}{2}+iz} - \Big(\tfrac{1}{2} - iz\Big) \mathcal{S}'_{jk}(z) Y^{-\frac{1}{2}-iz}\Big]^- \Big\}$$

$$= \frac{a^{2\tau}}{2} \sum_{k=1}^{n} \{2i\sigma\, \delta_{jk} Y^{-2\tau} - 2i\sigma |\mathcal{S}'_{jk}(z)|^2 Y^{2\tau}$$

$$+ 4i\tau\, \delta_{jk} \operatorname{Im}(\overline{\mathcal{S}}'_{jk}(z) Y^{2i\sigma})\} \ .$$

We recall that $\mathcal{S}'(\sigma)$ is a unitary matrix for each σ so that

$$\sum_k |\mathcal{S}'_{jk}(\sigma)|^2 = 1 \qquad \text{for all } j \ .$$

Making use of this fact, dividing through by $-4\sigma\tau i$ and taking the limit as $\tau \to 0$, we obtain (9.32) after a short calculation. This completes the proof of Lemma 9.2. ■

PROOF OF LEMMA 9.1. It is clear from the definitions of e^Y and e_Y that in the kth cusp neighborhood

$$(9.42) \qquad e_{jk}^Y - e_{jk,Y} = \begin{cases} e_{jk}^a = e_{j2,k} - e_{j2,k}^{(0)} & \text{for } y > Y \\ 0 & \text{otherwise} . \end{cases}$$

Consequently

$$\|e_j^Y\|^2 - \|e_{j,Y}\|^2 = \sum_{k=1}^{n} \int_{y>Y} |e_{jk}^a|^2 \frac{dx\, dy}{y^2}$$

tends to zero as Y goes to infinity for each σ. Also it follows from (9.32) and (9.42) that

$$0 \le \|e_j^Y\|^2 - \|e_{j,Y}\|^2 \le \sum_{j=1}^{n} \|e_j^a\|^2 \le \text{const.} + \frac{i}{2}\,\mathrm{tr}(\mathcal{S}'^*(\sigma)\partial_\sigma \mathcal{S}'(\sigma)) .$$

As we have noted before the second term on the right in this inequality is the rate of increase of the winding number $M(\sigma)$ of the $\det \mathcal{S}'(\sigma)$, as asserted in (9.30). This completes the proof of Lemma 9.1. ∎

We turn next to the task of determining the trace of $T_Y C T_Y$. To this end we construct a different representation for the operator $U(t)$.

Consider the solution of the wave equation

$$u_{tt} = y^2 \Delta u + \frac{1}{4} u \tag{9.43}$$

defined over the entire Poincaré plane Π. Denote by $U_0(t)$ the operator relating initial data of solutions of (9.43) to their data at time t. In what follows we need consider only initial data $\{u(0), u_t(0)\}$ for which $u(0) = 0$.

Let Γ be a discontinuous subgroup of $G = SL(2,\mathbb{R})/\{\pm 1\}$ and F some fundamental domain for Γ. Since the Laplace-Beltrami operator appearing on the right in (9.43) is invariant under the action of G, it follows that solutions of the wave equation (9.43) are invariant under G. Moreover since the solutions of the wave equation are uniquely determined by their initial data, it follows that $U_0(t)$ and the transformations γ in G commute. We conclude from this that if the initial data of a solution $u(w,t)$ are automorphic with respect to Γ, then u is automorphic with respect to Γ for all t.

Denote by f initial data $\{0, f\}$ which is zero outside the fundamental domain F. Then

$$\sum_{\gamma \in \Gamma} \gamma f \tag{9.44}$$

is automorphic with respect to Γ and the data at time t of the solution with initial data (9.44) is

(9.45)
$$\sum_{\gamma \in \Gamma} U_0(t)\gamma f \, .$$

The restriction of (9.45) to F gives $U(t)f$; that is

(9.46)
$$U(t) = \chi_F \sum_{\gamma \in \Gamma} U_0(t)\gamma$$

for data in $\mathcal{H}_2$.

We now construct an explicit solution for the wave equation (9.43) in the Poincaré plane Π by the method of descent from the corresponding three dimensional equation:

$(9.43)_3$
$$v_{tt} = y^2(\partial_{x_1}^2 v + \partial_{x_2}^2 v) + y^3\partial_y(y^{-1}v_y) + v$$

in the half space: $\{w = (x_1, x_2, y);\ y > 0\}$.

It is readily verified that equation $(9.43)_3$ has spherical wave solutions of the form:

(9.47)
$$v(w,t) = h(r \pm t)/\sinh r \ ,$$

where h is an arbitrary function and r denotes the noneuclidean distance between the two points w and w':

(9.48)
$$\cosh r = 1 + \frac{|w-w'|^2}{2yy'} \ ;$$

here $|w-w'|$ is the ordinary euclidean distance between w and w'. Clearly the function obtained by a superposition of spherical waves, that is

$$\frac{1}{4\pi}\int f(w')\,\frac{h(r-t)}{\sinh r}\,\frac{dx_1'\,dx_2'\,dy'}{(y')^3}$$

is a solution of $(9.43)_3$. Taking h to be a delta function, we obtain the solution

(9.49) $$v(w,t) = \frac{1}{4\pi} \int_{r=|t|} \frac{f(w')}{\sinh t} \frac{dS'}{(y')^2} \equiv K_3(t)f .$$

It is easy to show that

$$\lim_{t \to 0} K_3(t)f = 0 \quad \text{and} \quad \lim_{t \to 0} \partial_t K_3(t)f = f .$$

We remark, although we do not need it here, that since $\partial_t K_3(t)f$ is also a solution of the wave equation, the solution to the general initial value problem:

$$v(w,0) = f_1 \quad \text{and} \quad v_t(w,0) = f_2 ,$$

is simply

$$v(w,t) = \partial_t K_3(t) f_1 + K_3(t) f_2 .$$

This incidently shows that the *solutions of* $(9.43)_3$ *satisfy Huygens' principle*.

The method of descent is in this case complicated by the fact that a solution v of $(9.43)_3$ which is independent of x_2 is *not* a solution of (9.43). However this difficulty is easily remedied; a simple calculation shows that if v satisfies $(9.43)_3$, then

$$u = y^{-1/2} v$$

satisfies the equation

$$u_{tt} = y^2 \Delta u + \frac{1}{4} u .$$

With this substitution, formula (9.49) becomes

$$u(w,t) = \frac{1}{4\pi} y^{-1/2} \int_{r=|t|} \frac{(y')^{1/2} f(w')}{\sinh t} \frac{dS'}{(y')^2} .$$

Finally if f is independent of x_2 the integral on the right side can be written as an integral over the disk $\{r < |t|\}$:

$$u(x,y,t) = \frac{\operatorname{sgn} t}{2^{3/2}\pi} \int_{r<|t|} \frac{f(x',y')}{\sqrt{\cosh t - \cosh r}} \frac{dx'\,dy'}{(y')^2} \equiv K_2(t)f \; ; \tag{9.50}$$

alternatively we can express this as the integrand over Π of the real part of the integrand. This is the explicit formula for the solution of the wave equation (9.43) with initial data

$$u(x,y,0) = 0 \quad \text{and} \quad u_t(x,y,0) = f(x,y) \; .$$

Combining this with (9.46) we get the following formula for solutions automorphic with respect to Γ:

$$u_t(x,y,t) = \frac{\operatorname{sgn} t}{2^{3/2}\pi} \sum_{\gamma\in\Gamma} \partial_t \int \operatorname{Re}\left(\frac{f(x',y')}{\sqrt{\cosh t - \cosh r_\gamma}}\right) \frac{dx'\,dy'}{(y')^2} \; , \tag{9.51}$$

where r_γ denotes the noneuclidean distance from the point (x',y') to the point $\gamma(x,y)$. The expression (9.51), being an even function of t, also represents the second component of

$$\tfrac{1}{2}(U(t) + U(-t))f \; .$$

Multiplying (9.51) by $\phi(t)$ and integrating gives, according to (9.21), the operator C; it is convenient to integrate the resulting expression by parts with respect to t so as to throw the differentiation onto ϕ instead of K_2. To apply the truncation operator T_Y we simply cut down the domain of integration to $F_0(Y)$. In this way we obtain as the kernel of $T_Y C T_Y$:

$$-\chi_{F_0(Y)}(x,y) \frac{1}{2^{3/2}\pi} \int \phi'(t) \operatorname{sgn} t \sum_{\gamma\in\Gamma} \operatorname{Re}\left(\frac{1}{\sqrt{\cosh t - \cosh r_\gamma}}\right) \chi_{F_0(Y)}(x',y')\,dt \; .$$

Note that if $\phi(t)$ has compact support, a convenient assumption to start with, then the sum inside the t integration is finite; the number of non-zero terms, does, however, depend on Y.

It is easy to verify for smooth ϕ with compact support that the kernel of $T_Y C T_Y$ is continuous on $F_0(Y) \times F_0(Y)$. It follows by property (v) of trace class operators that for such ϕ the trace is given by formula (9.5) as the integral of its kernel along the diagonal; that is

$$(9.52)\quad \operatorname{tr}(T_Y C T_Y) = \frac{-1}{2^{3/2}\pi} \int \left\{ \phi'(t) \operatorname{sgn} t \sum_{\gamma \in \Gamma} \int_{F_0(Y)} \operatorname{Re}\left(\frac{dx\, dy}{y^2 \sqrt{\cosh t - \cosh r_\gamma}} \right) \right\} dt$$

where in this expression r_γ denotes the noneuclidean distance from (x, y) to $\gamma(x, y)$.

We now proceed to derive the Selberg trace formula for ϕ with compact support. Later we shall show how to extend this result to the class of ϕ's limited only by conditions (9.14) and (9.18).

Following Selberg we partition the sum in (9.52) into groups of terms corresponding to the conjugacy classes of Γ. These subsums over the conjugacy classes are in turn grouped according to the type of transformation involved in the conjugacy class; these types are classified as identity, hyperbolic, elliptic and parabolic. We denote the sums over the conjugacy classes of these various types as

$$c(\mathrm{id}, Y),\quad c(\mathrm{hyp}, Y),\quad c(\mathrm{ell}, Y) \quad \text{and} \quad c(\mathrm{par}, Y)\ ,$$

respectively.

We shall show later on that as $Y \to \infty$

$$c(\mathrm{par}, Y) = n\phi(0) \log Y + \text{const.} + o(1)\ .$$

Comparing this with (9.33) we see that

$$(9.53)\qquad \lim_{Y \to \infty}\ [c(\mathrm{id}, Y) + c(\mathrm{hyp}, Y) + c(\mathrm{ell}, Y)]$$

exists. Since this is true for all ϕ, it holds in particular for ϕ replaced by a majorizing function ϕ_1 of compact support, satisfying (9.18) and such that

$$|\phi'(t)| \le -(\operatorname{sgn} t)\phi_1'(t) .$$

For such a ϕ_1 it is clear that all of the terms in (9.52) are positive and monotonically increasing in Y. It therefore follows by majorized convergence that we can compute the limit (9.53) as the sum of the limits as $Y \to \infty$ of the relevant terms in (9.52), that is, these terms with F(Y) replaced by F. This we now proceed to do.

For γ = identity, $r_\gamma = 0$ and the corresponding term in (9.52) becomes

$$c(\mathrm{id}, \infty) = -\frac{1}{2^{3/2}\pi}\left(\int \phi'(t)\,\frac{\operatorname{sgn} t}{\sqrt{\cosh t - 1}}\,dt\right) \text{ area } F \tag{9.54}$$

$$= -\frac{1}{4\pi}\left(\int \frac{\phi'(t)}{\sinh t/2}\,dt\right) \text{ area } F .$$

More generally let $\{\tau\}$ denote the conjugacy class of a given τ in Γ and let Γ_τ denote the centralizer of τ in Γ. Then it is easy to see that the terms of

$$\sum_{\gamma\in\{\tau\}} \cdot \frac{\operatorname{sgn} t}{2^{3/2}\pi} \int_F \operatorname{Re}\left(\frac{1}{\sqrt{\cosh t - \cosh r_\gamma}}\right)\frac{dx\,dy}{y^2} \tag{9.55}$$

are of the form

$$\frac{\operatorname{sgn} t}{2^{3/2}\pi} \int_F \operatorname{Re}\left(\frac{1}{\sqrt{\cosh t - \cosh r_{\alpha\tau\alpha^{-1}}}}\right)\frac{dx\,dy}{y^2} ;$$

introducing $\alpha^{-1}w$ as new variable, this becomes

$$\frac{\operatorname{sgn} t}{2^{3/2}\pi} \int_{\alpha^{-1}F} \operatorname{Re}\left(\frac{1}{\sqrt{\cosh t - \cosh r_\tau}}\right) \frac{dx\,dy}{y^2} .$$

Hence (9.55) can be rewritten as

$$\text{(9.55)}' \qquad \frac{\operatorname{sgn} t}{2^{3/2}\pi} \int_{F(\tau)} \operatorname{Re}\left(\frac{1}{\sqrt{\cosh t - \cosh r_\tau}}\right) \frac{dx\,dy}{y^2} ,$$

where

$$F(\tau) = \sum_{\alpha} \alpha^{-1} F ,$$

α running over a complete set of representatives from the left cosets $\Gamma \backslash \Gamma_\tau$.

We claim that $F(\tau)$ is a fundamental domain for Γ_τ. To see this we use the fact that F is a fundamental domain for Γ; this means that for any w in Π there is a γ in Γ which maps w into $\overline{F}$:

$$\gamma w \in \overline{F} .$$

Further there is some α in the same coset as γ such that

$$\gamma = \alpha\kappa , \qquad \kappa \in \Gamma_\tau .$$

Hence, since

$$\gamma w = \alpha\kappa w \in \overline{F} ,$$

we have

$$\kappa w \in \alpha^{-1}\overline{F} \subset \overline{F(\tau)} .$$

Similarly one can show that no $\kappa \neq e$ in Γ_τ maps an interior point of $F(\tau)$ into $F(\tau)$. This proves our contention. The expression (9.55)′ remains unchanged if we replace τ by $\sigma = g^{-1}\tau g$, g in G, and $F(\tau)$ by any fundamental domain for $(g^{-1}\Gamma_\tau g)_\sigma$.

For a discrete subgroup Γ, every centralizer Γ_τ is a cyclic group consisting of the transformations in Γ with the same fixed points as τ. If τ is hyperbolic or parabolic then Γ_τ is infinite and if τ is elliptic Γ_τ is finite. A generator of Γ_τ is called a *primitive element*; every element of Γ is a power of some primitive element which is uniquely determined up to an inversion. In what follows we shall for definiteness choose a particular primitive element for each centralizer in Γ. Note that the centralizer of any nontrivial element of Γ_τ is again Γ_τ so that $F(\tau^k) = F(\tau)$ if $\tau^k \neq$ identity.

Suppose first that τ is a chosen primitive hyperbolic element of Γ. In the calculation of (9.55)′ we replace τ by the conjugate transformation

$$\sigma = \begin{pmatrix} \rho^{1/2} & 0 \\ 0 & \rho^{-1/2} \end{pmatrix}, \qquad \rho > 1 ;$$

$\rho(\tau)$ is called the norm of τ. We choose for $F(\sigma)$ the fundamental domain

$$F(\sigma) = [(x, y) : 1 < y < \rho] .$$

We now compute the contribution to the trace formula of the conjugacy class $\{\tau^k\}$; in this case (9.55)′ becomes

$$(9.56) \qquad \frac{\operatorname{sgn} t}{2^{3/2}\pi} \int_{F(\sigma)} \operatorname{Re}\left(\frac{1}{\sqrt{\cosh t - \cosh r_{\sigma^k}}}\right) \frac{dx\,dy}{y^2} .$$

Now

$$\cosh r_{\sigma^k} = 1 + \frac{(x^2 + y^2)(\rho^k - 1)^2}{2\rho^k y^2} \leq \cosh t$$

if and only if

$$\left(\frac{x}{y}\right)^2 \leq \frac{2\rho^k \cosh t - \rho^{2k} - 1}{(\rho^k - 1)^2} .$$

Hence (9.56) vanishes if

$$\cosh t \le \frac{\rho^{2k}+1}{2\rho^k} \;;$$

and if we set $x = sy$, (9.56) is transformed into

$$(9.56)' \qquad \frac{\operatorname{sgn} t}{2\pi} \iint_1^{\rho} \frac{\rho^{k/2}}{\sqrt{2\rho^k \cosh t - (\rho^{2k}+1) - s^2(\rho^k-1)^2}} \frac{dy}{y}\, ds \;,$$

here the s integration is over the interval

$$s^2 \le \frac{2\rho^k \cosh t - \rho^{2k} - 1}{(\rho^k-1)^2} \;.$$

(9.56)′ is readily evaluated as

$$\frac{\rho^{k/2} \log \rho}{2|\rho^k - 1|} \operatorname{sgn} t$$

and the contribution of the conjugacy class $\{\gamma^k\}$ to $c(\mathrm{hyp}, \infty)$ is

$$-\frac{\rho^{k/2}\log\rho}{2|\rho^k-1|} \int\limits_{\cosh t > \frac{\rho^{2k}+1}{2\rho^k}} \phi'(t) \operatorname{sgn} t\, dt = \frac{\rho^{k/2}\log\rho}{2|\rho^k-1|}\left[\phi\left(\cosh^{-1}\left(\frac{\rho^{2k}+1}{2\rho^k}\right)\right) + \phi\left(-\cosh^{-1}\left(\frac{\rho^{2k}+1}{2\rho^k}\right)\right)\right].$$

Replacing ρ by $\exp \ell$ and making use of the fact that ϕ is even, this takes on a somewhat simpler form:

$$\frac{\ell}{2|\sin h\, k\ell|} \phi(k\ell) \;.$$

The total contribution of the hyperbolic elements in Γ to (9.52) is therefore

$$(9.57)\qquad c(\text{hyp},\infty) = \sum_{\substack{\text{chosen hyp.}\\ \text{primitives } \tau}} \sum_{k\neq 0} \frac{\ell(\tau)}{2\left|\sinh \frac{k\ell(\tau)}{2}\right|} \phi(k\ell(\tau)) .$$

The elliptic contribution can be treated similarly. A primitive elliptic transformation τ in Γ is conjugate in G to a transformation of the form:

$$\sigma = \begin{pmatrix} \cos\omega & -\sin\omega \\ \sin\omega & \cos\omega \end{pmatrix},$$

where $\omega = \pi/m$ for some integer $m > 1$. $F(\sigma)$ can be taken as a sector with geodesic sides, vertex at $(0,1)$ and angle ω. The action of σ^k, $k = 1,2,\cdots,m-1$, leaves invariant circles centered in the noneuclidean sense about $(0,1)$. If w is a point of noneuclidean distance $\log\rho$ from $(0,1)$ then

$$\cosh r_{\sigma^k} = \frac{2\rho^2 + (\rho^2-1)^2 \sin^2 k\omega}{2\rho^2} .$$

The double integral (9.55) becomes decoupled if we introduce "polar" coordinates:

$$x = \frac{\sin\theta\cos\theta(\rho^2-1)}{\cos^2\theta + \rho^2\sin^2\theta}, \qquad y = \frac{\rho}{\cos^2\theta + \rho^2\sin^2\theta} .$$

In this case

$$\frac{dx\,dy}{y^2} = \frac{(\rho^2-1)}{\rho^2}\, d\rho\, d\theta$$

and

$$F(\sigma) = \{(\rho,\theta);\ 1 < \rho < \rho_0,\ |\theta| < \omega/2\} ,$$

where ρ_0 is the root of the equation $r_{\sigma^k} = |t|$ which is > 1. If we now replace ρ by

$$s = \frac{2\rho^2 + (\rho^2-1)^2 \sin^2 k\omega}{2\rho^2} ,$$

the integral (9.55)′ becomes simply

$$\frac{\omega \operatorname{sgn} t}{4\pi \sin k\omega} \int_1^{\cosh t} \frac{1}{[(\cosh t-s)(s-1+2\sin^2 k\omega)]^{1/2}}\, ds$$

$$= \frac{\omega \operatorname{sgn} t}{4\pi \sin k\omega}\left[\frac{\pi}{2} - \sin^{-1}\left(\frac{\sin^2 k\omega - \sinh^2 t/2}{\sin^2 k\omega + \sinh^2 t/2}\right)\right].$$

Taking the time derivative of this expression, we see that the contribution of the conjugacy class $\{\tau^k\}$ to the trace formula is

$$\frac{\omega}{4\pi}\int_{-\infty}^{\infty} \phi(t)\, \frac{\cosh t/2}{\sinh^2 t/2 + \sin^2 k\omega}\, dt$$

and hence that the total elliptic contribution is

$$(9.58) \qquad c(\mathrm{ell}, \infty) = \sum_{\substack{\text{chosen ell.} \\ \text{primitive } \tau}} \ \sum_{k=1}^{m(\tau)-1} \int_{-\infty}^{\infty} \phi(t)\, \frac{\cosh t/2}{\sinh^2 t/2 + \sin^2 k\omega(\tau)}\, dt\,.$$

Finally we treat the parabolic contribution to the trace formula. If the fundamental domain for Γ has n inequivalent cusps, then Γ contains precisely n inconjugate centralizers of parabolic type. As in (8.1) each of these is conjugate in G to the integral translation group Γ_∞ for which

$$\sigma = \begin{pmatrix} 1 & 1 \\ 0 & 1 \end{pmatrix}$$

is a primitive element and

$$F(\sigma) = \{(x, y);\ |x| < 1/2\}$$

is a fundamental domain.

In computing the contribution to (9.52) of the conjugacy class $\{\sigma^k\}$, the first step is to compute the terms of

$$\sum_{\gamma\in\{\sigma^k\}} \frac{\operatorname{sgn} t}{2^{3/2}\pi} \int_{F_0(Y)} \operatorname{Re}\left(\frac{1}{\sqrt{\cosh t - \cosh r_\gamma}}\right)\frac{dx\,dy}{} \tag{9.59}$$

If we proceed as in (9.55) with F replaced by $F_0(Y)$, we see that a typical term of (9.59) is of the form

$$I_k(\alpha, Y, t) = \frac{\operatorname{sgn} t}{2^{3/2}\pi} \int_{F_0(Y)} \operatorname{Re}\left(\frac{1}{\sqrt{\cosh t - \cosh r_{\alpha^{-1}\sigma^k\alpha}}}\right)\frac{dx\,dy}{y^2}$$

$$= \frac{\operatorname{sgn} t}{2^{3/2}\pi} \int_{\alpha^{-1}F_0(Y)} \operatorname{Re}\left(\frac{1}{\sqrt{\cosh t - \cosh r_{\sigma^k}}}\right)\frac{dx\,dy}{y^2} .$$

Notice that

$$\cosh r_{\sigma^k} = 1 + \frac{k^2}{2y^2} \le \cosh t$$

if and only if

$$y > \frac{|k|}{2|\sinh t/2|} \equiv b_k(t) . \tag{9.60}$$

We may suppose that α has been chosen from a left coset of Γ_∞ so that $\alpha^{-1}F \subset F(\sigma)$; in this case the $\alpha^{-1}F$ are disjoint and fill out $F(\sigma)$. As $Y \to \infty$, $I_k(\alpha, Y, t)$ converges monotonically to $I_k(\alpha, \infty, t)$ and in fact it would be convenient to replace $I_k(\alpha, Y, t)$ by $I_k(\alpha, \infty, t)$ for all but the identity α. The sum of the differences

$$\sum_{\alpha\neq \mathrm{id}} [I_k(\alpha, \infty, t) - I_k(\alpha, Y, t)] \tag*{(9.61)$_k$}$$

is majorized for $b_k(t) < a$ by

$$\frac{1}{2^{3/2}\pi}\int_{b_k}^{a}\frac{1}{\sqrt{\cosh t - 1 - \dfrac{k^2}{2y^2}}}\frac{dx\,dy}{y^2} = \frac{1}{2\pi|k|}\int_1^{a/b_k}\frac{1}{\sqrt{s^2-1}}\frac{ds}{s}$$

$$= \frac{1}{2\pi|k|}\cos^{-1}\frac{b_k}{a} \le \frac{1}{4|k|}$$

and by zero for $a < b_k(t)$. Since $b_k(t) < a$ only if $|k| < 2a|\sinh t/2|$, the expressions $(9.61)_k$, summed over $k \neq 0$, is less than const. $\log|\sinh t/2|$; that is, it is of the order of $|t|$. Multiplying this by $\phi'(t)$, of compact support, the result is absolutely integrable. It therefore follows by majorized convergence that we can make the above replacement in (9.59) and work instead with

$$I_k(Y,t) \equiv I_k(\mathrm{id}, Y, t) + \sum_{\alpha \neq \mathrm{id}} I_k(\alpha, \infty, t)$$

$$= \frac{\operatorname{sgn} t}{2\pi|k|}\int_1^{Y/b_k}\frac{1}{\sqrt{s^2-1}}\frac{ds}{s}$$

for $b_k(t) < Y$ and equal to zero otherwise. Differentiating this expression with respect to t we get

$$\partial_t I_k(Y,t) = \begin{cases} \dfrac{1}{4\pi|k|}\dfrac{1}{\sqrt{\left(\dfrac{Y}{b_k}\right)^2 - 1}}\dfrac{\cosh t/2}{|\sinh t/2|} & \text{for } Y > b_k \\ 0 & \text{for } b_k > Y\,. \end{cases}$$

The contribution to the trace formula due to each parabolic cusp is therefore

$$\sum_{k\neq 0}\int \phi(t)\partial_t I_k(Y,t)\,dt = \sum_{k=1}^{\infty} 2\int \phi(t)\partial_t I_k(Y,t)\,dt\,.$$

In order to derive an estimate for this sum we follow Kubota [p. 103, 7] and use the familiar summation formula

$$(9.62)\qquad \sum_{k=1}^{\infty} f(k) = \frac{f(1)}{2} + \int_1^{\infty} f(x)\,dx + \int_1^{\infty} f'(x)\theta(x)\,dx\ ,$$

where θ denotes the saw-toothed function:

$$\theta(x) = x - [x] - 1/2\ .$$

In our case

$$f(x) = 2 \int_{|t| > 2\sinh^{-1} x/2Y} \phi(t)\partial_t I_x(Y,t)\,dt\ .$$

We now estimate successively the terms on the right side of (9.62), beginning with

$$(9.62)_1\quad \frac{f(1)}{2} = \frac{1}{4\pi} \int_{|t|>2\sinh^{-1} 1/2Y} \phi(t) \frac{1}{\sqrt{(2Y\sinh t/2)^2-1}} \frac{\cosh t/2}{|\sinh t/2|}\,dt$$

$$= \frac{1}{2\pi} \int_{|s|>1} \phi(2\sinh^{-1}(s/2Y)) \frac{ds}{s\sqrt{s^2-1}}$$

$$= \frac{1}{2}\phi(0) + o(1) \qquad \text{as } Y \to \infty\ .$$

Next we compute

$$(9.62)_2\quad \int_1^{\infty} f(x)\,dx = \frac{1}{2\pi} \int_1^{\infty} \int \phi(t) \frac{1}{\sqrt{(2Y\sinh t/2)^2-x^2}} \frac{\cosh t/2}{|\sinh t/2|}\,dt\,dx\ ,$$

where the range of integration for the inner integral is $|t| > 2\sinh^{-1}x/2Y$. Interchanging the order of integration and setting $x = 2Y|\sinh t/2|s$ and $\varepsilon = \sinh^{-1}(1/2Y)$, we obtain

$$\int_1^\infty f(x)\,dx = \frac{1}{2\pi}\int_{|t|>\varepsilon} \phi(t)\frac{\cosh t/2}{|\sinh t/2|}\int_{(2Y|\sinh t/2|)^{-1}}^{1} \frac{1}{\sqrt{1-s^2}}\,ds\,dt$$

$$= \frac{1}{2\pi}\int_{|t|>\varepsilon} \phi(t)\frac{\cosh t/2}{|\sinh t/2|}\left[\frac{\pi}{2} - \sin^{-1}(2Y|\sinh t/2|)^{-1}\right] dt$$

$$\equiv A + B\ .$$

The analysis for A can be read off from Kubota [p. 105, 12]:

$$A = \frac{1}{2}\int_\varepsilon^\infty \phi(t)\,\frac{\cosh t/2}{\sinh t/2}\,dt$$

$$= -\phi(0)\,e^{-\varepsilon}\log\varepsilon + \frac{1}{2}\int_\varepsilon^\infty \phi(t)\,dt + \phi(0)\int_\varepsilon^\infty e^{-t}\log t\,dt$$

$$- \frac{1}{2\pi}\int_{-\infty}^\infty \Phi(\sigma)\int_\varepsilon^\infty\left[\frac{e^{-t}}{t} - \frac{e^{-t(1+i\sigma)}}{1-e^{-t}}\right] dt\,d\sigma$$

$$= \phi(0)\log Y + \frac{1}{4}\,\Phi(0) - \phi(0)\,C - \frac{1}{2\pi}\int_{-\infty}^\infty \Phi(\sigma)\,\frac{\Gamma'(1+i\sigma)}{\Gamma(1+i\sigma)}\,d\sigma + o(1)\ ,$$

where C denotes Euler's constant. Moreover setting $s = (2Y\sinh t/2)^{-1}$, B becomes

$$B = \frac{-1}{\pi}\int_{-1}^1 \phi\left(2\sinh^{-1}\frac{1}{2sY}\right)\sin^{-1}s\,\frac{ds}{s} = -\phi(0)\log 2 + o(1)\ .$$

Finally we compute the third term in (9.62). For this we need a suitable expression for $f'(x)$. With the substitution

$$s = \frac{2Y \sinh t/2}{x}$$

we can rewrite $f(x)$ as

$$f(x) = \frac{2}{\pi x}\int_1^\infty \phi\Big(2 \sinh^{-1} \frac{xs}{2Y}\Big) \frac{ds}{s\sqrt{s^2-1}} .$$

Hence

$$f'(x) = -\frac{2}{\pi x^2}\int_1^\infty \phi\left(2 \sinh^{-1} \frac{xs}{2Y}\right) \frac{ds}{s\sqrt{s^2-1}}$$

$$+ \frac{4}{\pi x^2}\int_1^\infty \phi'\left(2 \sinh^{-1} \frac{xs}{2Y}\right) \frac{xs}{\sqrt{4Y^2+x^2s^2}} \frac{ds}{s\sqrt{s^2-1}} \equiv A' + B' .$$

It is clear that as $Y \to \infty$,

$$A' = -\frac{\phi(0)+o(1)}{x^2} = O\left(\frac{1}{x^2}\right)$$

and that

$$B' = \frac{o(1)}{x^2} = O\left(\frac{1}{x^2}\right) .$$

Since (cf. Kubota [p. 106, 12])

$$-\int_1^\infty \frac{\theta(x)}{x^2}\, dx = C - \frac{1}{2} ,$$

we obtain

$$(9.62)_3 \qquad \int_1^\infty f'(x)\theta(x)\, dx = \phi(0)(C-1/2) + o(1) .$$

Summarizing we see that the total parabolic contribution to the trace formula truncated at Y is

$$9.63)\quad n\left[\phi(0)\log Y-\phi(0)\log 2+\frac{1}{4}\Phi(0)-\frac{1}{2\pi}\int_{-\infty}^{\infty}\Phi(\sigma)\frac{\Gamma'(1+i\sigma)}{\Gamma(1+i\sigma)}\,d\sigma\right]+o(1).$$

Substituting the estimates (9.19), (9.33) and (9.63) into (9.23) and passing to the limit as $Y\to\infty$, we finally obtain the Selberg trace formula:

$$(9.64)\qquad \sum\Phi(i\lambda_j)+\eta\Phi(0)+\sum\Phi(\nu_j)$$

$$=c(\mathrm{id},\infty)+c(\mathrm{hyp},\infty)+c(\mathrm{ell},\infty)$$

$$-n\left\{\phi(0)\log 2-\frac{1}{4}\Phi(0)+\frac{1}{2\pi}\int_{-\infty}^{\infty}\Phi(\sigma)\frac{\Gamma'(1+i\sigma)}{\Gamma(1+i\sigma)}\,d\sigma\right\}$$

$$+\frac{n}{4}\Phi(0)-\frac{i}{4\pi}\int_{-\infty}^{\infty}\Phi(\sigma)\,\mathrm{tr}(\mathcal{S}'^*(\sigma)\partial_\sigma\mathcal{S}'(\sigma))\,d\sigma\ ;$$

here $c(\mathrm{id},\infty)$, $c(\mathrm{hyp},\infty)$ and $c(\mathrm{ell},\infty)$ are given in (9.54), (9.57) and (9.58) respectively. We note that by Proposition 8.14, $\mathrm{tr}\,\mathcal{S}'(0)=n$.

Having established the trace formula for smooth ϕ with compact support, it is easy to show that it holds for an extended class of functions.

THEOREM 9.4. *The Selberg trace formula is valid for all even real-valued functions satisfying the conditions* (9.14) *and* (9.18).

PROOF. For a given even real-valued ϕ satisfying (9.14) and (9.18) we construct an approximating sequence $\{\phi_k\}$ of smooth functions with compact support as follows: Choose a decreasing function ξ in $C^\infty[0,\infty]$ such that

$$\xi(s)=\begin{cases}1 & \text{for}\quad 0\le s<1\\ 0 & \text{for}\quad s>2\ ,\end{cases}$$

set

$$\xi_k(t) = \xi(|t|/k)$$

and define

$$\phi_k(t) = \xi_k(t)\phi(t) \ . \tag{9.65}$$

It is clear from (9.14) that

$$\Phi_k(\sigma) \to \Phi(\sigma) \quad \text{and} \quad \Phi_k'(\sigma) \to \Phi'(\sigma) \qquad \text{for all } \sigma \tag{9.66$_1$}$$

and it is not hard to show from (9.18)[1] that

---

[1] Taking the Fourier transform of ϕ_k as defined in (9.65) we obtain

$$\Phi_k(\sigma) = \int k\,\Xi(k\sigma')\,\Phi(\sigma-\sigma')\,d\sigma' ,$$

where

$$\Xi(\sigma) = 2\int_0^\infty \cos\sigma t\,\xi(t)\,dt \ .$$

Now Peetre's inequality asserts that

$$k(\sigma)^a \le 2^{|a|}k(\sigma')^a k(\sigma-\sigma')^{|a|}$$

where $k(\sigma) = (1+|\sigma|^2)^{1/2}$. It is clear that a similar inequality holds for $(1+|\sigma|)$. Hence setting

$$p(\sigma) = (1+|\sigma|)^{3+\delta}$$

we get

$$p(\sigma) \le \text{const.}\ p(\sigma')\,p(\sigma-\sigma') ,$$

from which it follows that

$$p(\sigma)|\Phi_k'(\sigma| \le \text{const.} \int (k\,p(\sigma')\,\Xi(k\sigma'))\,p(\sigma-\sigma')|\Phi'(\sigma-\sigma')|\,d\sigma' .$$

Making use of 9.18) and the obvious inequality

$$\int p\left(\frac{\sigma'}{k}\right)\Xi(\sigma')\,d\sigma' < \text{const.} ,$$

we obtain $(9.66)_2$.

$$(9.66)_2 \qquad |\Phi_k'(\sigma)| \leq \frac{\text{const.}}{(1+|\sigma|)^{3+\delta}} ,$$

uniformly in k.

We shall use the subscript k to denote the terms in (9.64) which correspond to ϕ_k. We wish to establish the convergence with respect to k of the various terms in (9.64) as well as the absolute convergence of the sums in c(hyp) and c(ell); we omit the by now superfluous ∞ from these symbols. We have previously shown in (9.17) that the sum on the left is absolutely convergent and equal to $\operatorname{tr} C_p$. Moreover an integration by parts as in (9.17) gives

$$\sum \Phi_k(\nu_j) = -\int_0^\infty \Phi'(\tau)N(\tau)\,d\tau$$

and it follows by majorized convergence from this, (9.16) and $(9.66)_{1,2}$ that

$$\lim_{k\to\infty} \operatorname{tr} C_{p,k} = \operatorname{tr} C_p .$$

A similar argument using (9.34) shows that

$$\int_{-\infty}^{\infty} \Phi_k(\sigma)\operatorname{tr}(\mathcal{S}'^*(\sigma)\partial_\sigma \mathcal{S}'(\sigma))\,d\sigma \to \int_{-\infty}^{\infty} \Phi(\sigma)\operatorname{tr}(\mathcal{S}'^*(\sigma)\partial_\sigma \mathcal{S}'(\sigma))\,d\sigma .$$

The limit

$$\int_{-\infty}^{\infty} \Phi_k(\sigma)\frac{\Gamma'(1+i\sigma)}{\Gamma(1+i\sigma)}\,d\sigma \to \int_{-\infty}^{\infty} \Phi(\sigma)\frac{\Gamma'(1+i\sigma)}{\Gamma(1+i\sigma)}\,d\sigma$$

also follows by majorized convergence using $(9.66)_1$,

$$\left|\frac{\Gamma'(1+i\sigma)}{\Gamma(1+i\sigma)}\right| \leq \text{const. } \log(2+|\sigma|)$$

and

$$|\Phi_k(\sigma)| \leq \frac{\text{const.}}{(1+|\sigma|)^{2+\delta}} \qquad \text{for all } k ,$$

which is a direct consequence of $(9.66)_2$.

Combining these various results we see that the limit as k tends to ∞ of

$$c_k(\text{id}) + c_k(\text{hyp}) + c_k(\text{ell})$$

exists. It remains to show that the limit of the sums in this expression is equal to the sum of the limits of the terms. It is clear from $(9.66)_{1,2}$ that

$$\phi_k'(t) \to \phi'(t)$$

boundedly and it follows directly from this and (9.54) that

$$c_k(\text{id}) \to c(\text{id}) .$$

We treat the hyperbolic and elliptic terms by a less direct argument. To this end choose an even real-valued ϕ^* satisfying (9.14), (9.18) and the inequality

$$|\phi(t)| \leq \phi^*(t) .$$

Applying the above arguments to ϕ^* we see that

$$\lim_{k\to\infty} [c_k^*(\text{hyp}) + c_k^*(\text{ell})]$$

exists. According to (9.57) and (9.58) the individual terms in the sums comprising $c_k^*(\text{hyp})$ and $c_k^*(\text{ell})$ are positive and monotonic increasing in k. We conclude from this that the corresponding sums for $c^*(\text{hyp})$ and $c^*(\text{ell})$ converge. Since the terms in these sums majorize the corresponding terms for $c_k(\text{hyp})$ and $c_k(\text{ell})$, it follows by majorized convergence that

$$\lim_{k\to\infty} [c_k(\text{hyp}) + c_k(\text{ell})] = \text{sum of terms in (9.57) and (9.58)}$$

and furthermore that the sum on the right is absolutely convergent. This completes the proof of Theorem 9.4. ∎

Appendix 1 to Section 9: Another trace formula

In this appendix we outline another way of obtaining a trace formula for discontinuous subgroups of G which is based on the semigroup of operators $\{Z(t)\}$ introduced in Section 2. It will be recalled that $Z(t)$ is defined in terms of the incoming and outgoing subspaces $\mathcal{D}_{\pm}$ which in turn depend on the choice of the cut-off point, $y = a$, in each cusp neighborhood. Fortunately the trace formula obtained from Z does not depend on a and the limiting procedure as $a \to \infty$, which was basic to the Selberg trace formula (9.64), is now used only as a convenience in evaluating some of the terms.

We shall have to deal with operators which are not symmetric. The notion of trace class is readily generalized to such operators.

DEFINITION. A bounded operator K defined on a Hilbert space is of *trace class* if the symmetric non-negative operator

$$(K^*K)^{1/2} \tag{9.67}$$

is of trace class in the sense of (9.1).

The analogues of properties (i)-(v) are valid for these more general trace class operators. In particular for any orthonormal basis $\{f_j\}$ the sum of the diagonal elements of the matrix representation for K,

$$\sum (Kf_j, f_j) , \tag{9.68}$$

converges, its value is independent of the choice of basis and the common value is again called the *trace of* K.

Since the eigenvectors of a nonsymmetric operator do not in general form a basis, the relation (9.3) does not follow, as before, directly from the definition of the trace. Nevertheless this relation is still valid:

THEOREM 9.5 (Lidskii). *The trace of a trace class operator* K *is equal to the sum of the eigenvalues of* K, *each eigenvalue being counted with multiplicity equal to the dimension of its eigenspace.*

For a proof of this important theorem we refer the reader to the book of Gohberg and Krein [11].

We shall apply these ideas to the operator

$$K = \int_0^\infty \phi(t)\, Z(t)\, dt \ , \tag{9.69}$$

where ϕ is C_0^∞ on $\mathbf{R}_+$ and $Z(t)$ is the semigroup of operators defined as in Appendix 2 to Section 3:

$$Z(t) = P_{\mathcal{D}} U(t) P_{\mathcal{D}} \ , \qquad t \geq 0 \ ; \tag{9.70}$$

$U(t)$ being the solution of the automorphic wave equation on $\mathcal{H}_G$ (Theorem 5.4) and $P_{\mathcal{D}}$ being the E-*orthogonal projection* which removes the $\mathcal{D}_-$ and $\mathcal{D}_+$ components; here $\mathcal{D}_\pm$ are defined as in (6.10). We see by (7.13)′ that $P_{\mathcal{D}}$ can be defined as follows: Setting $P_{\mathcal{D}} f = g$, it is clear that $g = f$ on $F_0(a)$ and that in the jth cusp neighborhood

$$g_{1j}(w) = f_{1j}(w) - f_{1j}^{(0)}(y) + \left(\frac{y}{a}\right)^{1/2} f_{1j}^{(0)}(a) \ , \tag{9.71}$$

$$g_{2j}(w) = f_{2j}(w) - f_{2j}^{(0)}(y) \ .$$

Note that since the E-form and the G-form are the same in the cusp neighborhoods for $y > a$ it follows that $P_{\mathcal{D}}$ is orthogonal with respect to both of these forms. We denote the infinitesimal generator of Z by B.

As before the trace formula will be obtained by equating two different expressions for the trace of K, one in terms of the eigenvalues of K and the other in terms of the kernel of K. The proof of this formula has been split up into four steps.

Step 1. The relation between the eigenfunctions of B *and the outgoing eigenfunctions of* A. As before we denote by $e = e(z) = e(x, y; z)$ an eigenfunction of A with eigenvalue iz. That is, e is a vector function with two components, both C^∞ and automorphic, and e satisfies the partial differential equation

$$Ae = ize .$$

In any cusp neighborhood we can by integrating this equation with respect to x obtain an ordinary differential equation for the zero Fourier component of e which in the jth cusp is denoted by $e_j^{(0)}$. This equation can be solved explicitly and we get for $z \neq 0$

$$e_j^{(0)} = a_j\{y^{1/2-iz}, iz\, y^{1/2-iz}\} + b_j\{y^{1/2+iz}, iz\, y^{1/2+iz}\} . \tag{9.72}$$

The eigenfunction e(z) is called *outgoing* if in each cusp neighborhood

a) $e - e_j^{(0)}$ has finite G-norm

and

b) $b_j = 0$.

It follows from Corollary 6.4 that an outgoing eigenfunction e is E-orthogonal to all data in $\mathfrak{D}_-$ which have compact support.

THEOREM 9.6. *There is a one-to-one correspondence between the eigenfunctions of* B *and the outgoing eigenfunctions of* A *provided the eigenvalue* $iz \neq 0$.

PROOF. Suppose that e(z) is an outgoing eigenfunction of A and define g by

$$g = P_{\mathfrak{D}} e , \tag{9.73}$$

where by the definition of $P_{\mathfrak{D}}$ in (9.71),

$$P_{\mathfrak{D}} e = e - e_{\mathfrak{D}} ; \tag{9.74}$$

$e_{\mathfrak{D}}$ vanishes off the cusp neighborhoods and in the jth cusp

$$e_{\mathcal{D}j} = \begin{cases} a_j\{y^{1/2-iz} - y^{1/2}a^{-iz}, iz\, y^{1/2-iz}\} & \text{for} \quad y > a \\ 0 & \text{for} \quad y < a\ . \end{cases}$$

Since $U(t)$ is the solution operator of a hyperbolic equation, it can be applied to data with local finite energy. In particular $U(t)e$ and $U(t)e_{\mathcal{D}}$ can be defined; in fact, they can be written down explicitly. On the one hand e satisfies $Ae = ize$; it follows that

$$U(t)e = e^{izt}e\ , \tag{9.75}$$

since both sides satisfy the differential equation $u_t = Au$ and have the same initial data. On the other hand, by $(6.9)_+$

$$U(t)e_{\mathcal{D}} = \{u(t), u_t(t)\}$$

where in the jth cusp neighborhood

$$u(y,t)_j = \begin{cases} y^{1/2-iz}e^{izt} - y^{1/2}a^{-iz} & \text{for} \quad y > ae^t \\ 0 & \text{for} \quad y < ae^t\ . \end{cases}$$

It follows from (9.71) that

$$P_{\mathcal{D}}U(t)e_{\mathcal{D}} = 0\ . \tag{9.76}$$

Using (9.73), (9.74), (9.75), (9.76) and the definition of Z gives

$$Z(t)g = P_{\mathcal{D}}U(t)P_{\mathcal{D}}g = P_{\mathcal{D}}U(t)P_{\mathcal{D}}e = P_{\mathcal{D}}U(t)e - P_{\mathcal{D}}U(t)e_{\mathcal{D}} = e^{izt}P_{\mathcal{D}}e = e^{izt}g\ .$$

Differentiating this with respect to t at $t = 0$, we finally obtain

$$Bg = izg\ . \tag{9.77}$$

This proves half of Theorem 9.6.

To prove the converse assertion, let g be an eigenfunction of B. From (9.77) we see that at each point in F, except on the lines $y = a$ in each cusp neighborhood, g satisfies the differential equation

$$Ag = izg \ .$$

In particular we see that in each cusp neighborhood, for $y < a$ the zero Fourier component $g_j^{(0)}$ of g is of the form (9.72). We claim that $b_j = 0$ in each cusp; if this is so we can define e(z) to be equal to g except that its zero Fourier component is equal to $a_j\{y^{1/2-iz}, iz\,y^{1/2-iz}\}$ throughout the jth cusp neighborhood for all j. It is clear that e(z) is then outgoing and is related to g by (9.73).

To show that $b_k = 0$ for the kth cusp, we choose τ so small that all points (x, y) which lie in the half strip

$$-1/2 < x < 1/2 \,, \quad y > a e^{-\tau}$$

belong to the kth cusp neighborhood. This is certainly possible if a is sufficiently large. We then choose a C_o^∞ function ξ so that

$$\xi(y) = \begin{cases} 0 & \text{for} \quad y < a \,, \\ 0 & \text{for} \quad y > a e^{\tau} \,. \end{cases} \tag{9.78}$$

We define the function $d = d_k$ as follows: $d = 0$ everywhere except in the kth cusp and in the kth cusp neighborhood

$$d = \{y^{1/2}\xi(y), \ y^{3/2}\xi'(y)\} \ . \tag{9.79}$$

It follows from the definition $(6.10)_-$ of $\mathcal{D}_-$ and from $\xi(y) = 0$ for $y < a$ that d belongs to $\mathcal{D}_-$.

Since points for which $y > a e^{-\tau}$ belong to the kth cusp neighborhood, it follows that formula $(6.9)_-$, describing the action of U(t), is applicable to d for $t = \tau$:

$$U(\tau)d = \{y^{1/2}\xi(ye^{\tau}), y^{3/2}e^{\tau}\xi'(ye^{\tau})\} \ \text{in the} \ k^{th} \ \text{cusp} \ . \tag{9.80}$$

Since by (9.78) $\xi(y) = 0$ for $y > ae^{\tau}$, it follows that $U(\tau)d = 0$ for $y > a$; therefore $U(\tau)d$ is E-orthogonal to $\mathfrak{D}$. Finally since g is an eigenfunction of B, it is also an eigenfunction of Z and in fact

$$Z(\tau)g = e^{iz\tau}g \ . \tag{9.81}$$

We can now write the following series of identities:

$$\begin{aligned} e^{iz\tau}E(g, U(\tau)d) &= E(Z(\tau)g, U(\tau)d) = E(P_{\mathfrak{D}}U(\tau)g, U(\tau)d) \\ &= E(U(\tau)g, U(\tau)d) = E(g, d) = 0 \ . \end{aligned} \tag{9.82}$$

In the first equality we have used (9.81), in the second equality the definition of Z, in the third equality the E-orthogonality of $U(\tau)d$ to $\mathfrak{D}$, in the fourth equality the E-isometry of U and in the fifth equality the fact that d belongs to $\mathfrak{D}_-$ and that g is E-orthogonal to $\mathfrak{D}_-$.

Since $U(\tau)d$, given by formula (9.80), is zero except in the kth cusp neighborhood where it is independent of x, we can write

$$E(g, U(\tau)d) = E(g_k^{(0)}, U(\tau)d) \ . \tag{9.83}$$

Now the function $g_k^{(0)}$ is of the form (9.72). Substituting this in the right side of the expression (9.80) we find that the coefficient of a_k is zero, that is

$$E(\{y^{1/2-iz}, iz\,y^{1/2-iz}\}, \{y^{1/2}\xi, y^{3/2}e^{\tau}\xi'\}) = 0 \ ; \tag{9.84}$$

this follows from the E-orthogonality of outgoing and incoming data. The coefficient of b_k is

$$2iz \int y^{iz} e^{\tau} \xi'(ye^{\tau})\,dy \ .$$

Obviously ξ can be chosen so that this quantity is equal to 1. Then it follows from this, (9.84) and (9.83) that

$$E(g, U(\tau)d) = b_k .$$

Using (9.82) we deduce that $b_k = 0$. This completes the proof of Theorem 9.6. ∎

The argument presented above can be carried over to eigenfunctions of B of higher *index*, that is to the nullspace of $(B-iz)^k$ for $k > 1$.

COROLLARY 9.7. *The relation* (9.73) *gives a one-to-one correspondence between elements of the null space of* $(B-iz)^k$ *and outgoing elements in the null space of* $(A-iz)^k$ *if the corresponding eigenvalue is not zero.*

LEMMA 9.8. A *has no outgoing eigenfunctions of index greater than one corresponding to purely imaginary eigenvalues not equal to zero.*

PROOF. We have to show for σ real $\neq 0$ and g outgoing, that

$$(A-i\sigma)^2 g = 0 \tag{9.85}$$

implies that $(A-i\sigma)g = 0$. Let us denote the left side of this relation by f:

$$(A-i\sigma)g = f . \tag{9.86}$$

We claim that $f = 0$. Now it is clear by (9.85) that

$$(A-i\sigma)f = 0 \tag{9.87}$$

and since A takes outgoing data into outgoing data, f is outgoing. It therefore follows from Theorem 8.5 that f is a proper eigenfunction, that is that f belongs to $\mathcal{H}_G$. The zero Fourier component of a solution of (9.87) has the form (9.72) in each cusp neighborhood. Hence in order for f to be L_2, both a and b have to be zero; that is in all cusp neighborhoods

$$f^{(0)} = \{0, 0\} . \tag{9.88}$$

We denote the components of f by f_1, f_2 and those of g by g_1, g_2. Recalling the definition (3.4) of A, we can rewrite equation (9.87) componentwise as follows:

(9.89)
$$\begin{aligned} f_2 - i\sigma f_1 &= 0 \\ Lf_1 - i\sigma f_2 &= 0 . \end{aligned}$$

Eliminating f_2 we obtain

$$(L+\sigma^2)f_1 = 0 .$$

It then follows from the first equation in (9.89) that $(L+\sigma^2)f_2 = 0$. Combining the last two relations we obtain

(9.90)
$$(L+\sigma^2)(f_2+i\sigma f_1) = 0 .$$

Next we write equation (9.86) in component form:

$$\begin{aligned} g_2 - i\sigma g_1 &= f_1 \\ Lg_1 - i\sigma g_2 &= f_2 . \end{aligned}$$

Eliminating g_2 we get

(9.91)
$$(L+\sigma^2)g_1 = f_2 + i\sigma f_1 .$$

Denote by F(Y) the fundamental domain F minus that portion of each cusp neighborhood for which $y > Y$. By Green's formula

$$\iint_{F(Y)} \bar{g}_1 (L+\sigma^2)(f_2+i\sigma f_1)\,\frac{dx\,dy}{y^2} = \iint_{F(Y)} \overline{(L+\sigma^2)g_1}\,(f_2+i\sigma f_1)\,\frac{dx\,dy}{y^2} + B(Y) ,$$

where B(Y) is the boundary term

$$B(Y) = \sum_j \int_{y=Y} [\bar{g}_{1j}\partial_y(f_2+i\sigma f_1)_j - (\partial_y \bar{g}_{1j})(f_2+i\sigma f_1)_j]\,dx .$$

Using the relation (9.90) on the left and (9.91) on the right we get

$$\iint_{F(Y)} |f_2 + i\sigma f_1|^2 \frac{dx\,dy}{y^2} = -B(Y) .$$

By assumption $f - f^{(0)}$ and $g - g^{(0)}$ have finite energy; furthermore according to (9.88) $f^{(0)} = 0$. It follows from this that there is a subsequence of Y's tending to infinity such that B(Y) tends to 0. This proves that the L_2 norm of $f_2 + i\sigma f_1$ over F is zero and so

$$f_2 + i\sigma f_1 = 0 .$$

Combining this with the first relation in (9.89) we conclude for $\sigma \neq 0$ that $f_1 = 0$ and $f_2 = 0$. In other words, $f = 0$ and substituting this into (9.86), we see that the proof of Lemma 9.8 is complete. ∎

According to Theorem 3.2, if $-\sigma^2$ belongs to the point spectrum of L then $\pm i\sigma$ belongs to the point spectrum of A. We denote the collection of these as $\pm i\nu_j$. According to Theorem 9.6, $i\nu_j$ also belong to the point spectrum of B and according to Corollary 9.7 and Lemma 9.8 these eigenvalues have index 1 for B.

As in (3.11), denote by λ_j^2, $j = 1, \cdots, m$, the positive eigenvalues of L, $\lambda_j > 0$. Then the functions f_j^+, defined in $(3.12)_+$, are L_2 outgoing eigenfunctions of A with eigenvalue λ_j. Moreover f_j^-, defined by $(3.12)_-$, will be an outgoing eigenfunction of A with eigenvalue $-\lambda_j$ whenever f_j^- is E-orthogonal to $\mathcal{D}_-$; these values of λ_j were called *nonrelevant* in the definition stated before Theorem 3.9.

One can show: *The eigenvalues* λ_j *and* $-\lambda_j$, $j = 1, 2, \cdots, m$, *described above, have index one.* The proof of this assertion is similar to but simpler than that of Lemma 9.8 since all of the eigenfunctions are assumed to lie in $\mathcal{H}_G$. We leave the details to the reader.

In addition to the proper eigenfunctions of A with eigenvalues on the imaginary and real axes described above, A has generalized eigenfunctions e(z) of the kind constructed abstractly in Section 2 and given concrete realization in Sections 7 and 8. The zero Fourier components of these eigenfunctions in the various cusp neighborhoods is given by (8.27). Taking the residue of an $e_k(z)$ at a pole $z = -i\mu$ of $\mathcal{S}'(z)$, we obtain another eigenfunction of A and it is clear from (8.27) that such an eigenfunction will be outgoing since the residue of the incoming part of $e_k(z)$ vanishes. In view of Corollary 3.21 and Theorem 9.6, we expect the resulting outgoing eigenfunction of A to correspond to a μ-eigenvector of B. As defined in the second appendix to Section 3, an eigenvector f of B, with eigenvalue μ, is called a μ-eigenvector if $\operatorname{Re} \mu < 0$, if f is E-orthogonal to $\mathcal{H}_p'$ and if f does not belong to $\mathcal{P}_-$.

We now analyze the class of generalized (not proper) outgoing eigenfunctions of A and show that under the mapping

$$f = P_{\mathcal{D}} e \ , \tag{9.92}$$

these correspond to the μ-eigenvectors of B, provided the corresponding eigenvalue is different from zero.

For an outgoing eigenfunction e(z) of A, the zero Fourier coefficient in the cusp neighborhoods will be of the form (9.72) with all of the b_j but perhaps not all of the a_j equal to zero. If $\operatorname{Im} z < 0$, then e(z) will be bounded in all of the cusp neighborhoods and hence e(z) will be a proper eigenfunction. As shown above this is also the case when $\operatorname{Im} z = 0$ provided $z \neq 0$. Disregarding the case $z = 0$, it follows that if e(z) is not proper, then $\operatorname{Im} z > 0$.

It is clear that $f = \mathcal{P}_{\mathcal{D}} e(z)$ is an eigenvector of B:

$$Bf = izf \ .$$

If e(z) is outgoing but not proper and $z \neq 0$, then $\operatorname{Im} z > 0$ and not all of the a_j vanish. Consequently f will be discontinuous at $y = a$ in some cusp neighborhood and therefore f cannot belong to $\mathcal{P}_-$.

Finally we must show that f is E-orthogonal to $\mathcal{H}_p'$ and the null vectors of A. We note first of all that as in (2.33)

$$Bf = P_{\mathfrak{D}} Ae$$

and hence $Bf = Ae$ except for the zero Fourier coefficient in the cusp neighborhoods. If g is an eigenfunction in $\mathcal{H}'_p$, $Ag = i\sigma g$, then its zero Fourier coefficients vanish in the cusp neighborhoods. Thus when integrating a product of f and g, we can replace f by e and disregard the bad behavior of the zero Fourier coefficient of e. An integration by parts therefore gives:

$$E(Bf, g) + E(f, Ag) = 0 .$$

It follows that

$$iz\,E(f, g) = E(Bf, g) = -E(f, Ag) = i\sigma\,E(f, g)$$

and since $\operatorname{Im} z > 0$ we conclude that $E(f, g) = 0$. On the other hand, if g is a proper null vector of A then, by Theorem 5.4, g is a null vector of the form E and therefore $E(f, g) = 0$ in this case also.

We have now proved that any generalized eigenfunction of A, with eigenvalue different from zero, corresponds under the mapping (9.92) to a μ-eigenvector of B. According to Theorem 9.6 this mapping defines a one-to-one correspondence between the outgoing eigenfunctions of A and the eigenvectors of B, again excluding the case $z = 0$. Since all of the proper eigenfunctions and null vectors of A are otherwise accounted for, the *outgoing generalized eigenfunctions*, omitting null vectors, *are mapped by* (9.92) *onto the set of* μ*-eigenvectors of* B.

We turn now to the zero eigenvalue and the corresponding nullvectors of B, B^2, etc. Again these are related to the outgoing nullvectors of A, A^2, etc. by the relation (9.73). To study the nullvectors of A, $A^2, \cdots$ we write:

$$f = \{f_1, f_2\}, \quad Af = \{f_2, Lf_1\}, \quad A^2 f = \{Lf_1, Lf_2\}, \cdots .$$

It follows from these formulas that

$$Af = 0 \text{ if and only if } f_2 = 0 \text{ and } Lf_1 = 0 ;$$

$$A^2 f = 0 \text{ if and only if } Lf_1 = 0 \text{ and } Lf_2 = 0 .$$

Thus the nullspaces of A and A^2 can be built out of the nullspace of the scalar operator L. We denote the nullspace of L by $\mathcal{Q}$ and its elements by q. From the equation $Lq = 0$ we deduce that the zero Fourier component of q in each cusp neighborhood satisfies the ordinary differential equation

$$y^2 \partial_y^2 q^{(0)} + \frac{1}{4} q^{(0)} = 0 ,$$

whose solution is

$$q^{(0)} = a y^{1/2} + b y^{1/2} \log y . \tag{9.93}$$

We first consider the subspace $\mathcal{Q}_1$ of those solutions q for which $b = 0$ in (9.93) for all cusp neighborhoods. We denote the dimension of $\mathcal{Q}_1$ by η_1. It is clear that every outgoing nullvector of A is of the form

$$f = \{q, 0\}, \qquad q \text{ in } \mathcal{Q}_1 .$$

For if q were of the form (9.93) with $b_j \neq 0$ in, say, the jth cusp neighborhood, then in order to be outgoing, $f_2^{(0)}$ would have to be nonzero in this cusp and of the form

$$f_{2j}^{(0)} = -b_j y^{1/2} .$$

When q is in $\mathcal{Q}_1$, f belongs to $\mathcal{K}$ and is therefore also a nullvector of B. Conversely we claim that all nullvectors of B are of this form. For suppose that f is a nullvector of B; then $Bf = 0$ and $Z(t)f = f$ for all $t > 0$. Since $Bf = 0$, we conclude that $f_2 = 0$ and $Lf_1 = 0$ except on $y = a$ in the various cusp neighborhoods. The zero Fourier component of f_1 in each cusp neighborhood is of the form

$$a y^{1/2} + b y^{1/2} \log y \qquad \text{for } y < a$$

$$a y^{1/2} + b y^{1/2} \log a \qquad \text{for } y > a .$$

It is easy to verify with the aid of (6.9) that for small $t > 0$ the zero Fourier component of $U(t)f = \{u(t), u_t(t)\}$ in each cusp splits into incoming and outgoing parts and is of the form

$$u(t)^{(0)} = a y^{1/2} + \frac{b}{2} y^{1/2} \log a + \frac{b}{2} y^{1/2} \log(y e^{-t}) \quad \text{for } a e^{-t} < y < a .$$

It is clear from this that if $b \neq 0$, then $Z(t)f = P_{\mathcal{D}} U(t) f$ depends on t and hence that $Bf \neq 0$. Thus the class of outgoing nullvectors of A is the same as the class of nullvectors of B.

Next we consider the subspace $\mathcal{Q}_2$ of $\mathcal{Q}_1$ consisting of solutions q for which both a and b in (9.93) are equal to zero in all of the cusp neighborhoods; such q are in $L_2(F)$. Denote the dimension of $\mathcal{Q}_2$ by η_2; in the first part of Section 9, η was used for this number. Clearly every f of the form

$$f = \{0, q\}, \qquad q \text{ in } \mathcal{Q}_2 .$$

is an outgoing nullvector of A^2. Such an f belongs to $\mathcal{K}$ and therefore is also a nullvector of B^2.

It is *a priori* possible for there to exist other outgoing nullvectors of A^k and corresponding nullvectors of B^k for $k \geq 2$. We shall eventually prove that there are no such nullvectors. However for the time being we denote the dimension of the subspace spanned by such assorted nullvectors by η_3.

We summarize the above information about the spectrum of B: *The spectrum of* B *consists of*

i) $\{\pm i\nu_j\}$ where $-\nu_j^2$ belongs to the point spectrum of L;

ii) $\{\lambda_j\}$ where λ_j^2 belongs to the point spectrum of L, $\lambda_j > 0$;

iii) $\{-\lambda_j\}$ where λ_j is nonrelevant;

iv) $\{\mu_j\}$ where μ_j is the eigenvalue of an outgoing generalized eigenfunction of A;

v) 0 with the multiplicity $\eta_1 + \eta_2 + \eta_3$.

If σ is an eigenvalue of B then $e^{\sigma t}$ is the eigenvalue of $Z(t)$ corresponding to the same eigenvector. The corresponding eigenvalue of K, defined by (9.69) as $\int \phi(t)\, Z(t)\, dt$ is then

$$\int_0^\infty \phi(t) e^{\sigma t} dt = \Phi(-i\sigma)$$

where Φ denotes the Fourier transform of ϕ. It follows from the above list of eigenvalues of B that the *sum of the eigenvalues of* K *is*

$$\sum \Phi(\nu_j) + \sum \Phi(-\nu_j) + \sum \Phi(-i\lambda_j) + \sum{}' \Phi(i\lambda_j) + \sum \Phi(-i\mu_j) + (\eta_1 + \eta_2 + \eta_3)\, \Phi(0) . \tag{9.94}$$

The Σ' in the fourth sum means that we sum over the nonrelevant λ_j's.

Note that this sum depends only on the outgoing eigenfunctions of A and is therefore *independent of the choice of the cutoff value* a which enters into the definition of $\mathcal{D}_-$, $\mathcal{D}_+$ and Z.

Step 2. K *is of trace class.* The fact that K is of trace class is essential to this treatment. We have devised two different approaches to this problem. The first, which we present below, uses estimates already developed in the body of Section 9 and is therefore shorter. The second method, which is given in a second appendix, is more direct and straightforward.

We begin by introducing the operator:

$$C = \int_0^\infty \phi(t)\, U(t)\, dt .$$

We recall that P′ and Q′ are the E-orthogonal projections on $\mathcal{P}$ and $\mathcal{H}'_G$ respectively; $P' + Q' = I$. Hence

$$K = P_{\mathcal{D}} C P_{\mathcal{D}} = P_{\mathcal{D}} Q'CQ'P_{\mathcal{D}} + P_{\mathcal{D}} Q'CP'P_{\mathcal{D}} + P_{\mathcal{D}} P'CQ'P_{\mathcal{D}} + P_{\mathcal{D}} P'CP'P_{\mathcal{D}} .$$

Since P' has a finite dimensional range, so do all but the first of the operators on the right and consequently all but the first are of trace class.

It remains to prove that $P_{\mathcal{D}} Q'CQ'P_{\mathcal{D}}$ is of trace class. Now $Q'CQ'$ is determined by the action of U on $\mathcal{H}'_G$. We recall that the null space $\mathcal{J}$ of A is a subspace of $\mathcal{H}'_G$ of finite dimension by Theorem 5.2. Hence if $P_{\mathcal{J}}$ denotes the G-orthogonal projection of $\mathcal{H}'_G$ onto the G-orthogonal complement of $\mathcal{J}$, then as above we can reduce our problem to showing that

$$P_{\mathcal{D}} Q'P_{\mathcal{J}} CP_{\mathcal{J}} Q'P_{\mathcal{D}}$$

is of trace class.

Now $P_{\mathcal{J}}\mathcal{H}'_G$ is equivalent to $\mathcal{H}'_G/\mathcal{J}$ which in turn is equivalent to $\mathcal{H}'_E$ by Corollary 5.3. Consequently we can work with $Q'P_{\mathcal{J}} UP_{\mathcal{J}} Q'$ in the E-norm or, equivalently, with U on $\mathcal{H}'_E$. In particular the spectral theory developed for U on $\mathcal{H}'_E$ is applicable. It will be recalled that $\mathcal{H}'_E$ can be decomposed into two invariant subspaces corresponding to the point and continuous parts of the spectrum of A on $\mathcal{H}'_E$:

$$\mathcal{H}'_E = \mathcal{H}'_p + \mathcal{H}'_c .$$

We denote the corresponding parts of $Q'P_{\mathcal{J}} CP_{\mathcal{J}} Q'$ by C'_p and C'_c, respectively.

By repeating the argument which was used in (9.17) for C_p, one can prove that C'_p is of trace class in the E-norm and hence in the G-norm. It follows by trace class property (iii) that $P_{\mathcal{D}} C'_p P_{\mathcal{D}}$ is also of trace class.

As for C'_c, we see from Theorem 8.1 that it is an integral operator; in fact

$$G(C'_c f, f) = \sum_j \frac{1}{2\pi} \int \Phi(\sigma)\, E(f, e_j(\sigma))\, G(e_j(\sigma), f)\, d\sigma .$$

Consequently we can write

$$(9.95)\qquad G(P_{\mathcal{D}}C_c'P_{\mathcal{D}}f,f) = \sum_j \frac{1}{2\pi}\int \Phi(\sigma)\,E(f,P_{\mathcal{D}}e_j(\sigma))\,G(P_{\mathcal{D}}e_j(\sigma),f)\,d\sigma\ .$$

According to property (i), listed at the beginning of Section 9, the operator $P_{\mathcal{D}}C_c'P_{\mathcal{D}}$ will be of trace class if the sum

$$(9.96)\qquad \sum_k G(P_{\mathcal{D}}C_c'P_{\mathcal{D}}f_k, f_k)$$

converges for every G-orthonormal basis $\{f_k\}$. Since the E-form is continuous with respect to the G-form, it follows from (9.95) that the sum (9.96) converges absolutely if

$$(9.97)\qquad \int |\Phi(\sigma)|\,G(P_{\mathcal{D}}e_j(\sigma))\,d\sigma$$

is finite. An integration by parts as in the proof of Lemma 9.2 and the relation (9.32) can be used to show that

$$E(P_{\mathcal{D}}e_j(\sigma)) = 2\|e_j^a(\sigma)\|^2 + \frac{1}{\sigma}\,\mathrm{Im}\,(\overline{\mathcal{S}_{jj}'}(\sigma)\,a^{2i\sigma})\ .$$

Since

$$G(f) = E(f) + 2\iint_{F_0} \frac{|f_1|^2}{y^2}\,dx\,dy\ ,$$

we see that

$$(9.98)\qquad G(P_{\mathcal{D}}e_j(\sigma)) \le 4\|e_j^a(\sigma)\|^2 + \frac{1}{\sigma}\,|\mathrm{Im}(\overline{\mathcal{S}_{jj}'}(\sigma)\,a^{2i\sigma})|\ .$$

The finiteness of (9.97) now follows as in the proof of Lemma 9.3 from (9.34) and (9.98). We have now proved

THEOREM 9.9. K *is of trace class.*

It now follows by the Lidskii Theorem 9.5 and (9.94) that

$$\text{(9.99)}\qquad \operatorname{tr} K = \sum \Phi(\nu_j) + \sum \Phi(-\nu_j) + \sum \Phi(-i\lambda_j) + \sum{}' \Phi(i\lambda_j) + \sum \Phi(-i\mu_j) + (\eta_1 + \eta_2 + \eta_3)\Phi(0) .$$

The trace formula itself is obtained by equating (9.99) with a direct evaluation of $\operatorname{tr} K$ from an integral representation of K.

The construction of an integral representation of K proceeds as before. Combining (9.46) and (9.50) we see that the solution of the automorphic wave equation with initial data:

$$u(w,0) = 0 \quad \text{and} \quad u_t(w,0) = f ,$$

can be expressed in terms of an integral operator:

$$u(w,t) = R_F(t) f ,$$

with kernel

$$\text{(9.100)}\qquad R_F(w,w',t) = \frac{1}{2^{3/2}\pi} \sum_{\gamma \in \Gamma} \operatorname{Re}\left(\frac{1}{\sqrt{\cosh t - \cosh r_\gamma}}\right), \qquad t \geq 0 ,$$

here again r_γ denotes the non-Euclidean distance from w' to γw. It follows that the solution of the full initial value problem,

$$u(w,0) = f_1 \quad \text{and} \quad u_t(w,0) = f_2 ,$$

is given by

$$u(w,t) = \partial_t R_F(t) f_1 + R_F(t) f_2 ;$$

and hence that the operator $U(t)$ has the matrix kernel

$$\text{(9.101)}\qquad \begin{pmatrix} \partial_t R_F & R_F \\ \partial_t^2 R_F & \partial_t R_F \end{pmatrix} .$$

We recall that

$$C = \int \phi(t)\,U(t)\,dt ,$$

where ϕ belongs to $C_0^\infty(R_+)$. After an integration by parts we can therefore express the matrix kernel for C as

$$(9.102)\qquad \begin{pmatrix} -\int \phi'(t)\,R_F(t)\,dt & \int \phi(t)\,R_F(t)\,dt \\ \int \phi''(t)\,R_F(t)\,dt & -\int \phi'(t)\,R_F(t)\,dt \end{pmatrix} .$$

Finally K is obtained from C by operating on both sides with $P_{\mathcal{D}}$:

$$K = P_{\mathcal{D}}\, C P_{\mathcal{D}} = (K_{ij}) ;$$

the projection operator $P_{\mathcal{D}}$ has already been described in (9.71). Denoting the action of $P_{\mathcal{D}}$, on the first and second components of $\mathcal{H}_G$ for a given choice of a by T_1^a and T_2^a, respectively, we can write

$$(9.103)\qquad K_{ij} = T_i^a\, C_{ij}\, T_j^a .$$

Two properties of integral operators of the form

$$K_{12} = T_1^a \int \phi(t)\,R_F(t)\,dt\; T_2^a$$

are immediately evident. In the first place

$$g_1 = K_{12}\, f_2$$

lies in the first component space of $\mathcal{H}_G$ and hence by (4.26) and (5.7) the zero Fourier coefficient of g_1 in, say, the jth cusp neighborhood, when evaluated at $y = a$, is well defined; in fact a simple estimate gives

$$(9.104) \qquad |g_{1j}^{(0)}(a)| \leq \text{const. } G^{1/2}(\{g_1, 0\}) \leq \text{const. } \|f_2\| .$$

In the second place, since K is of trace class, so is each of its matrix elements; in particular this holds for K_{12}. It follows by (5.7) that the mapping defined as T_2^a from the first component of $\mathcal{H}_G$ to the second component is bounded and since $T_2^a T_1^a = T_2^a$, the operator

$$T_2^a \int \phi(t) R(t) dt \, T_2^a ,$$

considered as an operator on $L_2(F)$ is of trace class. We note that since all of the C_{ij} are of the form C_{12}, with ϕ replaced by derivatives of ϕ, the above considerations hold for all of them.

Step 3. Change of Space. The usual method for evaluating the trace of an integral operator is to integrate the kernel of the operator along the diagonal. However this requires that the operator act on an L_2 space. Unfortunately the operator K with which we have to deal acts on $\mathcal{H}_G$ whose first component is definitely not $L_2(F)$. The purpose of Step 3 is to put K into an L_2 framework.

It is fairly obvious, on account of the action of the operator T_1^a, that we may restrict the action of K to those elements of $\mathcal{H}_G$ whose first component is the sum of an L_2 function plus the functions $b_j y^{1/2}$ in the cusp neighborhoods. This suggests that we replace $\mathcal{H}_G$ by the space

$$(9.105) \qquad \mathcal{H}' = [L_2(F) \times C^n] \times L_2(F)$$

and the operator K by K′ defined below. We start by defining K'_{11}, the first of the four components of K′. For a given

$$f_1' = [f_1, b_j] \quad \text{in} \quad L_2(F) \times C^n$$

we set

$$K_{11}' f_1' = g' = [g, c_j]$$

where

(9.106) $$g = T_2^a C_d h \quad \text{and} \quad c_j = g_j^{(0)}(a-) ,$$

the function h and the operator C_d being defined as follows: $h = f_1$ except for its zero Fourier coefficient in the cusp neighborhoods when $y > a$; there

(9.107) $$h_j^{(0)} = b_j \left(\frac{y}{a}\right)^{1/2} .$$

To define the operator C_d we note that according to formula (9.102) the operators C_{11} and C_{22} on the diagonal of C are identical except that C_{11} acts on functions which are the first components of vectors in $\mathcal{H}_G$ whereas C_{22} acts on the second component. It follows that these operators can be extended to functions which are sums of the first component of a vector in $\mathcal{H}_G$ and the second component of another. We denote this extended operator by C_d. Clearly the function h defined in (9.107) can be so expressed and therefore h belongs to the domain of C_d.

This completes the definition of K_{11}'. We define K_{22}' as K_{22}; K_{12}' and K_{21}' are defined analogously.

The previous analysis shows that K_{22} as an operator on $L_2(F)$ to $L_2(F)$ is of trace class. It is clear that the operator K_{11}' defined above when restricted to $[L_2(F) \times 0]$ differs from K_{22} by a finite dimensional operator. It follows from this that K_{22}' is also of trace class. A similar analysis serves for the off diagonal components of K'.

We claim that K and K' have the same eigenfunctions. For it is clear that if f is an eigenfunction of K, then

$$f_1' = [T_2^a f_1, b_j] \text{ with } b_j = f_{1j}^{(0)}(a) \text{ and } f_2' = T_2^a f_2$$

is an eigenfunction of K'. Conversely, it is easy to see that every eigenfunction of K' is obtained in this fashion from an eigenfunction of K. It therefore follows from the Lidskii theorem that

$$\text{tr } K' = \text{tr } K \ . \tag{9.108}$$

Step 4. An explicit calculation of tr K'. It is clear that

$$\text{tr } K' = \sum_{i=1}^{2} \text{tr } K'_{ii} \tag{9.109}$$

and further that

$$\text{tr } K'_{11} = \text{tr } K'_{22} + \sum_{k=1}^{n} (K'_{11}\{0,\delta_{kj}\},\{0,\delta_{kj}\}) \ . \tag{9.110}$$

We first evaluate the second term on the right in (9.110); that is we have to determine the zero Fourier coefficient of $K'_{11}\{0,\delta_{kj}\}$ at $y = a$ in the jth cusp neighborhood. We have, setting $f'_1 = [0,\delta_{kj}]$ into (9.107) that

$$h(y) = \eta(y)\left(\frac{y}{a}\right)^{1/2}, \quad \text{where } \eta(y) = \begin{cases} 0 & \text{for } y < a \\ 1 & \text{for } y > a \ . \end{cases}$$

To be able to apply the operator C_d to this h we have to decompose it as the sum of a function in the domain of C_{11} and of one in the domain of C_{22}. Taking a C^∞ function ζ such that

$$\zeta(y) = \begin{cases} 0 & \text{for} \quad y \le a \\ 1 & \text{for} \quad y \ge a+1 \ , \end{cases}$$

we set

$$h = h_1 + h_2\,, \quad h_1 = \zeta h\,, \quad h_2 = (1-\zeta)h \ .$$

Then

$$C_d h = C_{11} h_1 + C_{22} h_2 \ .$$

To evaluate these operators we have first to find

$$U(t)\{h_1, 0\} \quad \text{and} \quad U(t)\{0, h_2\} . \tag{9.111}$$

This task is easy if a is sufficiently large compared to the lower end a_o of the jth cusp. Specifically suppose that

$$a > a_o e^T .$$

In this case it takes at least time T for any signal starting at $y = a$ to reach the end of the cusp. Therefore for $0 \leq t \leq T$ the solutions (9.111) can be written down explicitly by means of formula (6.9) as waves in $\mathfrak{D}_+$ and $\mathfrak{D}_-$, propagating up and down respectively in the cusp. The $\mathfrak{D}_+$ components of these waves are zero for $y \leq a$ and therefore do not contribute to the value of $g(a-)$. The value of the $\mathfrak{D}_-$ components are easily calculated and are given below:

$$\begin{aligned} U(t)\{h_1, 0\} &= \frac{1}{2}\left(\frac{y}{a}\right)^{1/2} \{\zeta(y e^t), y e^t \zeta'(y e^t)\} + d_+ , \\ U(t)\{0, h_2\} &= \frac{1}{2}\left(\frac{y}{a}\right)^{1/2} \{H(y e^t), \eta(y e^t)[1-\zeta(y e^t)]\} + d_+ , \end{aligned} \tag{9.112}$$

where $H(y)$ is the function $\int_a^y [1-\zeta(v)] \frac{dv}{v}$ and d_+ denotes the outgoing components.

Suppose that $\phi(t) = 0$ for $t > T$. Then formulas (9.112) are valid for all t in the support of ϕ. Setting $y = a$ in (9.112) and taking the first component of the first and the second component of the second solution respectively in (9.112) we get

$$C_d h(a) = C_{11} h_1(a) + C_{22} h_2(a) = \frac{1}{2} \int_0^T \phi(t) \{\zeta(ae^t) + \eta(ae^t)[1 - \zeta(ae^t)]\} dt$$

$$= \frac{1}{2} \int_0^T \phi(t) \eta(ae^t) dt = \frac{1}{2} \int_0^\infty \phi(t) dt = \frac{1}{2} \Phi(0) .$$

By definition of c_j in (9.106) we see that

$$c_j = \frac{1}{2}\delta_{kj}\,\Phi(0)$$

and therefore the sum on the right in (9.110) is

$$\frac{n}{2}\,\Phi(0)\ , \tag{9.113}$$

provided that a has been chosen sufficiently large.

It follows from this and the fact that $\operatorname{tr} K$ is independent of a that $\operatorname{tr} K'_{22}$ does not depend on a if a is sufficiently large. In this circumstance

$$\operatorname{tr} K'_{22}(a) = \operatorname{tr} K'_{22}(Y)$$

for $Y > a$.

Next let T_Y denote the truncation operator defined on $L_2(F)$ by

$$T_Y f = \chi_{F_0(Y)} f\ ,$$

where as before $\chi_{F_0(Y)}$ is the characteristic function of the set $F_0(Y)$. Since T_2^a truncates only the zero Fourier coefficients at $y \geq a$ in the cusp neighborhoods, it is clear that T_Y and T_2^a commute, that $T_Y T_2^Y = T_Y$ and that

$$T_Y(T_2^Y - T_2^a) = T_2^Y - T_2^a\ . \tag{9.114}$$

Making use of property (iii), listed at the beginning of Section 9, we have

$$\operatorname{tr}[T_Y(K'_{22}(Y) - K'_{22}(a))T_Y] = \operatorname{tr}[T_Y(K'_{22}(Y) - K'_{22}(a))]\ .$$

Now by (9.114)

$$\begin{aligned} T_Y(K'_{22}(Y) - K'_{22}(a)) &= T_Y T_2^Y C_{22} T_2^Y - T_Y T_2^a C_{22} T_2^a \\ &= T_Y(T_2^Y - T_2^a)C_{22}T_2^Y + T_Y T_2^a C_{22}(T_2^Y - T_2^a) \\ &= (T_2^Y - T_2^a)C_{22}T_2^Y + T_Y T_2^a C_{22}(T_2^Y - T_2^a)\ . \end{aligned}$$

Again employing property (iii) and (9.114) we see that

$$\operatorname{tr}[T_Y T_2^a C_{22}(T_2^Y - T_2^a)] = \operatorname{tr}[T_2^a C_{22}(T_2^Y - T_2^a) T_Y]$$
$$= \operatorname{tr}[T_2^a C_{22}(T_2^Y - T_2^a)] .$$

Combining these relations we obtain

$$\operatorname{tr}[T_Y(K'_{22}(Y) - K'_{22}(a)) T_Y]$$
$$= \operatorname{tr}[(T_2^Y - T_2^a) C_{22} T_2^Y + T_2^a C_{22}(T_2^Y - T_2^a)]$$
$$= \operatorname{tr}[K'_{22}(Y) - K'_{22}(a)] = 0 .$$

Since

$$T_Y K'_{22}(Y) T_Y = T_Y C_{22} T_Y ,$$

we conclude that

$$\operatorname{tr} T_Y K'_{22}(a) T_Y = \operatorname{tr} T_Y C_{22} T_Y . \tag{9.115}$$

Making use of property (iv) for trace operators we can now assert that

$$\operatorname{tr} K'_{22}(a) = \lim_{Y \to \infty} \operatorname{tr} T_Y K'_{22}(a) T_Y = \lim_{Y \to \infty} \operatorname{tr} T_Y C_{22} T_Y . \tag{9.116}$$

We have already computed the right side of (9.116) in the body of Section 9. In fact it is clear from (9.52) that the contributions of ϕ to $\lim \operatorname{tr}(T_Y C_{22} T_Y)$ for $t > 0$ and for $t < 0$ are the same. As a consequence the contributions to $\lim \operatorname{tr}(T_Y C_{22} T_Y)$ of the various conjugacy classes will be exactly *half of the values listed in* (9.54), (9.57), (9.58) and (9.63). Note that since $\phi(0) = 0$, the $\log Y$ term in $c(\text{par}, Y)$ is absent. We therefore obtain

$$\operatorname{tr} K'_{22} = \frac{1}{2} [c(\text{id}, \infty) + c(\text{hyp}, \infty) + c(\text{ell}, \infty) + c(\text{par}, \infty)] . \tag{9.117}$$

The final form of our trace formula is obtained by combining the relations (9.99), (9.108), (9.109), (9.110), (9.113) and (9.117). The result is

$$\text{(9.118)}\quad \sum \Phi(\nu_j) + \sum \Phi(-\nu_j) + \sum \Phi(-i\lambda_j) + \sum{}' \Phi(i\lambda_j)$$

$$+ \sum \Phi(-i\mu_j) + (\eta_1 + \eta_2 + \eta_3)\Phi(0)$$

$$= \frac{n}{2}\,\Phi(0) + c(\mathrm{id},\infty) + c(\mathrm{hyp},\infty) + c(\mathrm{ell},\infty) + c(\mathrm{par},\infty)\ .$$

It is of interest to compare the two trace formulas (9.64) and (9.118). With this in mind we again take ϕ real-valued and with compact support on $\mathbf{R}_+$ and denote its even extension on $\mathbf{R}$ by ϕ_1:

$$\phi_1 = \begin{cases} \phi(t) & \text{for} \quad t \geq 0 \\ \phi(-t) & \text{for} \quad t < 0\ . \end{cases}$$

Then

$$\Phi_1(\sigma) = \int_{-\infty}^{\infty} \phi_1(t)\, e^{i\sigma t}\, dt = \Phi(\sigma) + \Phi(-\sigma)\ .$$

The function pair ϕ_1 and Φ_1 is suitable for formula (9.64). In terms of Φ, the left side of (9.64) then becomes

$$\text{(9.119)}\quad \sum \Phi(i\lambda_j) + \sum \Phi(-i\lambda_j) + 2\eta_2\Phi(0) + \sum \Phi(\nu_j) + \sum \Phi(-\nu_j)\ .$$

Subtracting (9.64) from (9.118) and making use of (9.119) we obtain

$$\text{(9.120)}\quad \sum \Phi(-i\mu_j) - \sum{}'' \Phi(i\lambda_j) + (\eta_1 - \eta_2 + \eta_3)\Phi(0)$$

$$= \frac{i}{4\pi} \int_{-\infty}^{\infty} (\Phi(\sigma) + \Phi(-\sigma))\, \mathrm{tr}\,(\mathcal{S}'^{*}(\sigma)\, \partial_\sigma \mathcal{S}'(\sigma))\, d\sigma\ ;$$

where Σ'' in the second sum on the left denotes summation over the relevant λ_j's.

Since $\mathcal{S}'$ is unitary on the real axis, the relation $\mathcal{S}'^{*} = \mathcal{S}'^{-1}$ holds there. Using the linear algebra identity

$$\operatorname{tr}[\mathcal{S}'^{-1}(\sigma)\partial_\sigma \mathcal{S}'(\sigma)] = [\det \mathcal{S}'(\sigma)]^{-1}\partial_\sigma \det \mathcal{S}'(\sigma)$$

we can rewrite the right side of (9.120) as

$$\frac{i}{4\pi}\int_{-\infty}^{\infty} (\Phi(\sigma) + \Phi(-\sigma))\frac{\partial_\sigma D'(\sigma)}{D'(\sigma)}\, d\sigma \ ;$$

here we have used the abbreviation

$$\det \mathcal{S}'(\sigma) = D'(\sigma) \ .$$

If we could deform the integral involving $\Phi(\sigma)$ upward and that involving $\Phi(-\sigma)$ downward to $\pm i\infty$, we could then equate the above integral to the sum of its residues in the upper and lower half planes. Since $\mathcal{S}'$ is unitary on the real axis, its values in the upper and lower half planes are related by the Schwarz reflection principle, so that the contributions to the residue sum from the two half planes is the same and is precisely the first two terms on the left in (9.120). If this were so, we could deduce that the third term on the left is zero; that is

$$\eta_1 - \eta_2 + \eta_3 = 0 \ .$$

Referring back to the discussion at the end of Step 1, it is clear that $\mathcal{Q}_2$ is a subset of $\mathcal{Q}_1$ and hence that $\eta_2 \leq \eta_1$. It therefore would follow that

(9.121) $$\eta_1 = \eta_2 \quad \text{and} \quad \eta_3 = 0 \ .$$

Substituting this into (9.118) we would get

THEOREM 9.10. *For* Φ *the Fourier transform of a real-valued function in* $C_c^\infty(\mathbf{R}_+)$ *we have*

$$(9.122)\quad \sum \Phi(\nu_j) + \sum \Phi(-\nu_j) + \sum \Phi(-i\lambda_j) + \sum{}' \Phi(i\lambda_j) + \sum \Phi(-i\mu_j) + 2\eta_2\Phi(0) = \frac{n}{2}\Phi(0) + c(\mathrm{id},\infty) + c(\mathrm{hyp},\infty) + c(\mathrm{ell},\infty) + c(\mathrm{par},\infty)\ ;$$

here Σ' *denotes summation over the nonrelevant* λ_j's.

We do not have enough information about the size of $\mathcal{S}'(z)$ to show that these contour deformations are legitimate. However we succeeded in giving a purely function analytic proof of the equality of the integral on the right in (9.120) to the sum of residues in terms of a mini-trace formula. In remainder of this appendix we derive this mini-trace formula.

THEOREM 9.11. *Let* $\mathcal{S}''$ *denote the scattering matrix corresponding to the incoming and outgoing subspaces* $\mathcal{D}''_-$ *and* $\mathcal{D}''_+$. *In the notation of the previous theorem, the following relation holds*:

$$(9.123)\quad \sum \Phi(-i\mu_j) + \sum{}'' \Phi(i\lambda_j) = \frac{i}{2\pi}\int_{-\infty}^{\infty} \Phi(\sigma)\, \mathrm{tr}(\mathcal{S}''^*(\sigma)\partial_\sigma \mathcal{S}''(\sigma))\, d\sigma\ ;$$

here Σ'' *denotes summation over the relevant* λ_j's.

PROOF. We begin with the semigroup of operators $Z''(t)$ defined on the subspace

$$(9.124)\quad \mathcal{K}'' = \mathcal{H}'_c \ominus (\mathcal{D}''_- \oplus \mathcal{D}''_+)\ ,$$

which is associated with the scattering matrix $\mathcal{S}''$. According to Lemma 3.20 the eigenvectors of the generator B'' of Z'' are of two kinds: 1) Those which are also eigenvectors of B_+; these are simple eigenvectors with eigenvalues $-\lambda_j$ for the relevant λ_j's. 2) Those which are also eigenvectors of B^0_+; by Theorem 3.15 these eigenvectors are in one-to-one correspondence with the μ-eigenvectors of B.

Arguing as before we can now conclude that the smoothed-out operator

$$K'' = \int_0^\infty \phi(t)\, Z''(t)\, dt \tag{9.125}$$

will have as its eigenvalues the numbers $\Phi(-i\mu_j)$ and the $\Phi(i\lambda_j)$ for the relevant λ_j's. We shall prove that K'' is of trace class from which it follows by Theorem 9.5 that

$$\operatorname{tr} K'' = \sum \Phi(-i\mu_j) + \sum{}'' \Phi(i\lambda_j) , \tag{9.126}$$

where each term is counted according to the multiplicity of the corresponding eigenspaces.

It is easy to prove the trace class property for K'' from the fact that K is of trace class, where K denotes the operator defined in (9.69). In fact according to the Remark 3.6, $\mathcal{D}''_{\pm}$ is of finite codimension in $\mathcal{D}_{\pm}$. Hence we can write

$$P_{\mathcal{D}_{\pm}} = P_{\mathcal{D}''_{\pm}} + Q^o_{\pm} \tag{9.127}$$

where $P_{\mathcal{D}}$ denotes the orthogonal projection onto the complement of $\mathcal{D}$ and $Q^o_-\,[Q^o_+]$ denotes the orthogonal projection on $\mathcal{D}_- \ominus \mathcal{D}''_-\,[\mathcal{D}_+ \ominus \mathcal{D}''_+]$; the $Q^o_{\pm}$ both have finite dimensional ranges. Setting

$$K''_0 = P_{\mathcal{D}''_+} \int \phi(t)\, U(t)\, dt\, P_{\mathcal{D}''_-}$$

and noting that

$$K = P_{\mathcal{D}_+} \int \phi(t)\, U(t)\, dt\, P_{\mathcal{D}_-} ,$$

it follows from (9.127) that

$$K''_0 = K + \text{three degenerate operators}$$

and is therefore of trace class. Moreover K_0'' is also of trace class on the quotient space $\mathcal{H}/\mathcal{G}$, treated in Section 6. Since K'' is the restriction of K_0'' on $\mathcal{H}/\mathcal{G}$ to $\mathcal{H}_c'$, it follows that K'' is also of trace class.

We now obtain a realization of K'' as an integral operator in the outgoing $\mathcal{D}_+''$ translation representation and compute $\operatorname{tr} K''$ in the usual way from the kernel of this operator. In this representation $\mathcal{D}_+''$ corresponds to $L_2(R_+, C^n)$, where n is the number of inequivalent cusps in F. In order to characterize $\mathcal{D}_-''$ in the $\mathcal{D}_+''$ representation we recall that $\mathcal{D}_-''$ corresponds to $L_2(R_-, C^n)$ in the $\mathcal{D}_-''$ translation representation and that S'' maps the $\mathcal{D}_-''$ representation into the $\mathcal{D}_+''$ representation. Consequently in the $\mathcal{D}_+''$ translation representation, $\mathcal{D}_-''$ corresponds to $S''L_2(R_-, C^n)$ and $\mathcal{K}''$ to the orthogonal complement of $S''L_2(R_-, C^n)$ in $L_2(R_-, C^n)$. If $Q_\pm$ denotes the orthogonal projection of $L_2(R)$ onto $L_2(R_\pm)$, then the orthogonal projection onto $\mathcal{D}_-''$ is simply

$$S''Q_-S''^{-1} .$$

The action of

$$C = \int \phi(t)\, U(t)\, dt$$

in this representation is given by

$$Ck(s) = \int \phi(t)\, k(s-t)\, dt = \int \phi(s-t)\, k(t)\, dt .$$

In other words C corresponds to convolution by ϕ; in symbols

$$C = \phi * .$$

It follows that the action of K'' is that of a two-sided Toeplitz operator:

$$K'' \sim Q_-\phi*(Q_- - S''Q_-S''^{-1}) .$$

Since S'' is unitary we can rewrite this expression as

$$K'' \sim Q_- \phi * S''(S''^* Q_- - Q_- S''^*) .$$

By causality

$$S'' Q_- = Q_- S'' Q_-$$

and taking adjoints this becomes

$$Q_- S''^* = Q_- S''^* Q_- .$$

Consequently

$$S''^* Q_- - Q_- S''^* = S''^* Q_- - Q_- S''^* Q_- = Q_+ S''^* Q_-$$

and

(9.128) $$K'' \sim Q_- \phi * S'' Q_+ S''^* Q_- .$$

We recall that S'' commutes with translation and is therefore a distribution-valued convolution operator in the translation representation; causality implies that it has its support in $\mathbf{R}_-$. It follows that $\phi * S''$ corresponds to convolution by a well behaved matrix-valued function which we denote by $S''\phi$. Similarly S''^* is a distribution-valued convolution operator in the translation representation with support in $\mathbf{R}_+$. Proceeding formally we see by (9.128) that the action of K'' on functions in $L_2(\mathbf{R}_-)$ is given by the kernel

$$K''(u,s) = \int_0^\infty (S''\phi)(u-r) S''^*(r-s)\, dr , \qquad u, s < 0 .$$

Consequently

$$\operatorname{tr} K'' = \operatorname{tr} \int_{-\infty}^0 \int_0^\infty (S''\phi)(u-r) S''^*(r-u)\, dr\, du .$$

Setting $v = r - u$, $w = r + u$, this becomes

(9.129) $$\operatorname{tr} K'' = \frac{1}{2} \operatorname{tr} \int_0^{\infty} \int_{-v}^{v} (S''\phi)(-v)\, S''^*(v)\, dw\, dv$$

$$= \operatorname{tr} \int_0^{\infty} v(S''\phi)(-v)\, S''^*(v)\, dv \ .$$

Now multiplication by v corresponds to differentiation of the Fourier transform with a factor $-i$. Hence making use of the Plancherel theorem (9.129) can be rewritten as

$$\operatorname{tr} K'' = \frac{-i}{2\pi} \operatorname{tr} \int_{-\infty}^{\infty} \Phi(\sigma) \mathcal{S}''(\sigma)\, \partial_\sigma \mathcal{S}''^*(\sigma)\, d\sigma \ .$$

Making use of the trace property: tr AB = tr BA and the fact that

$$\mathcal{S}''^*(\sigma)\, \partial_\sigma \mathcal{S}''(\sigma) + (\partial_\sigma \mathcal{S}''^*(\sigma)) \mathcal{S}''(\sigma) \equiv 0$$

we finally obtain

(9.130) $$\operatorname{tr} K'' = \frac{i}{2\pi} \int_{-\infty}^{\infty} \Phi(\sigma) \operatorname{tr}(\mathcal{S}''^*(\sigma)\, \partial_\sigma \mathcal{S}''(\sigma))\, d\sigma \ .$$

Combining this with (9.126) gives the desired expression (9.123).

The above considerations can easily be made rigorous. As we have already noted $S''\phi$ is smooth and in L_2. On the other hand $S''^*(s)$ need not be a function. However we can approximate the action of S''^* in the strong topology by convolution by $(S''^*) * \psi_\delta$, where ψ_δ is an approximation of the delta function with support in $\mathbf{R}_+$. The previous development holds for S''^* replaced by $(S''^*) * \psi_\delta$ since the kernel of the approximate K'' is continuous and the trace can be computed by integrating along the diagonal. Passing to the limit as ψ_δ converges to the delta function and making use of a variant of property (iv) for trace class operators, we obtain (9.123) in full generality.

The right hand side of (9.123) is expressed in terms of $\mathcal{S}''$ rather than $\mathcal{S}'$ which appears in the Selberg trace formula. This discrepancy is easily remedied by means of the relation (9.124): $\mathcal{S}'' = \mathcal{S}_+ \mathcal{S}' \mathcal{S}_-$. Using properties of the trace it is readily verified that

$$\text{(9.131)} \quad \operatorname{tr}(\mathcal{S}''^* \partial_\sigma \mathcal{S}'') = \operatorname{tr}(\mathcal{S}'^* \partial_\sigma \mathcal{S}') + \operatorname{tr}(\mathcal{S}_-^* \partial_\sigma \mathcal{S}_-) + \operatorname{tr}(\mathcal{S}_+^* \partial_\sigma \mathcal{S}_+) .$$

According to Theorem 3.9, $\mathcal{S}_-$ and $\mathcal{S}_+$ are Blaschke products with zeros at precisely the points $-i\lambda_j$ (and poles at the conjugate points) for the relevant λ_j's. Hence replacing the integrals involving $\mathcal{S}_-$ and $\mathcal{S}_+$ which now appear on the right side of (9.123) by virtue of (9.131) we obtain

$$\text{(9.132)} \qquad \sum \Phi(-i\mu_j) - \sum{}'' \Phi(i\lambda_j) = \frac{i}{2\pi} \int_{-\infty}^{\infty} \Phi(\sigma) \operatorname{tr}(\mathcal{S}'^*(\sigma)\partial_\sigma \mathcal{S}'(\sigma))\, d\sigma .$$

Comparing (9.120) with (9.132) we are now able to deduce (9.121) which we state as

COROLLARY 9.12. *The outgoing nullvectors of* A *and* A^2 *are of the form* $\{q, 0\}$ *and* $\{0, q\}$, *respectively, with* q *in* $\mathcal{Q}_2$. *There are no outgoing eigenvectors of* A *of higher index than* 2.

Now all proper nullvectors of A are both incoming and outgoing. Moreover for q in $\mathcal{Q}_2$, $q^{(0)}$ vanishes in all cusp neighborhoods and $E(f) = 0$ for all f in $\mathcal{J}$. It therefore follows from Lemma 7.21 and Corollary 9.12 that

COROLLARY 9.13. *In the case of the modular group the null space* $\mathcal{J}$ *is trivial.*

Appendix 2 to Section 9. On the trace class property for K.

In this appendix we give a simple, direct proof that the operator K, defined by (9.69) as

$$K = \int_0^\infty \phi(t)\, Z(t)\, dt$$

with ϕ in $C_0^\infty(\mathbf{R}_+)$, is of trace class. Specifically we show

THEOREM 9.14. a) K *is an integral operator; its kernel is* C^∞ *except for discontinuities along the lines* $y = a$ *in each cusp neighborhood;*
b) K *differs by a finite dimensional operator from an operator* K_0 *whose kernel in each cusp neighborhood belongs to the Schwartz class of functions; that is, any derivative of this kernel tends to zero faster than any power of* y *and* y' *as* y *or* y' *tends to infinity.*

The difference between the kernel of K and that of K_0 is itself C^∞ except for discontinuities along the lines $y = a$ in the cusp neighborhoods, hence the kernel of K_0 is piecewise C^∞ and in each cusp neighborhood belongs to the Schwartz class. It is not hard to show that such an operator belongs to trace class on $L_2(F)$. But then it follows that K, which differs from K_0 by a finite dimensional operator is also of trace class on $L_2(F)$.

We turn now to the proof of Theorem 9.14. By definition

$$Z(t) = P_{\mathfrak{D}}\, U(t)\, P_{\mathfrak{D}} ,$$

so that

$$K = P_{\mathfrak{D}} \int \phi(t)\, U(t)\, dt\, P_{\mathfrak{D}} . \tag{9.133}$$

$U(t)$ is the solution operator for automorphic data of the wave equation

$$u_{tt} = y^2 \Delta u + \frac{1}{4} u = Lu , \tag{9.134}$$

and as such maps Cauchy data at time 0 into Cauchy data at time t. It can be regarded as an integral operator with a distribution-valued matrix kernel. We denote this kernel as

$$R_F(t, x, y, x', y') .$$

This kernel can be expressed in terms of the so-called Riemann matrix of the equation (9.134) considered as an equation over the entire Poincaré plane Π. This Riemann matrix

$$R(t, x, y, x', y')$$

is then the matrix solution of the vector form of (9.134):

$$R_t = AR \tag{9.135}$$

with initial values

$$\delta(x-x', y-y')\begin{pmatrix} 1 & 0 \\ 0 & 1 \end{pmatrix} . \tag{9.136}$$

The relation between R_F and R is given by

$$R_F(t, x, y, x', y') = \sum_{\gamma \in \Gamma} R(t, x, y, \gamma\{x', y'\}) . \tag{9.137}$$

The properties of R, stated in the next lemma, are fairly close to the surface.

LEMMA 9.15.

a) $R = R(t, x-x', y, y')$;

b) *For any positive number* c,

$$R(t, cx, cy, cy') = R(t, x, y, y') ; \tag{9.138}$$

c) $R(t, x, y, y') = 0$ *when the non-Euclidean distance between* $\{x, y\}$ *and* $\{x', y'\}$ *exceeds* t.

PROOF. Part (a) asserts that R depends only on the difference of x and x'. This follows from two facts: Since the coefficients of equation (9.134) are independent of x, the x-translate of a solution:

$$R(t, x+k, y, x'+k, y')$$

is again a solution. Moreover the initial value of this translated solution is again (9.136). So the translated solution is identical with $R(t,x,y,x',y')$.

To prove part (b) we observe that the operator $L = y^2\Delta + 1/4$ commutes with dilation; that is if we set

$$v_c(x, y) = v(cx, cy) ,$$

then

$$Lv_c = (Lv)_c . \tag{9.139}$$

It follows that if $v(t, x, y)$ is a solution of (9.134) over Π then so is $v(t, cx, cy)$. Thus $R(t, cx, cy, cx', cy')$ is a solution of (9.134) over Π. The initial value of this solution is $\delta(c(x-x'), c(y-y'))$ which, with respect to the measure $dx\, dy/y^2$ is equal to $\delta(x-x', y-y')$. This proves that R is invariant under dilation.

To prove part (c) we observe that the non-Euclidean sound speed for the equation (9.134) is one. Therefore a signal which originates at $\{x',y'\}$ cannot reach the point $\{x, y\}$ in time t if the distance between the two points exceeds t. ∎

LEMMA 9.16. *We define a generalized energy* E_2 *as follows*:

$$E_2(f) = ((L-I)^{-2}f_2, (L-I)^{-2}f_2) - ((L-I)^{-2}f_1, (L-I)^{-1}f_1 \tag{9.140}$$

where the bracket $(\,,)$ *denotes the* L_2 *scalar product over* Π. *Let* u *be any solution of* (9.134) *and denote its generalized energy at time* t *by* $E_2(t)$. *Then the following energy estimate holds*:

$$E_2(t) \le e^{4t/3}E(0) \qquad \textit{for } t > 0 . \tag{9.141}$$

PROOF. We rewrite equation (9.134) as

$$u_{tt} - (L-I)u = u .$$

Then we apply $(L-I)^{-2}$ to both sides and form the L_2 scalar product with $(L-I)^{-2}u_t$. This gives the identity:

$$(9.142)\ ((L-I)^{-2}u_t,(L-I)^{-2}u_{tt}) - ((L-I)^{-2}u_t,(L-I)^{-1}u = ((L-I)^{-2}u_t,(L-I)^{-2}u).$$

We estimate the right side by Schwarz's inequality as follows:

$$(9.143) \qquad ((L-I)^{-2}u_t,(L-I)^{-2}u) \le \|(L-I)^{-2}u_t\|\ \|(L-I)^{-2}u\|$$

$$\le \frac{1}{2}\|(L-I)^{-2}u_t\|^2 + \frac{1}{2}\|(L-I)^{-2}u\|^2 .$$

We recall from Section 4, see (4.8), that the operator $L-1/4$ is negative over Π; it follows from this that

$$\|(L-I)^{-1}u\|^2 \le -4/3(u,(L-I)^{-1}u) .$$

This shows that the right side of (9.143) is less than or equal to

$$\frac{1}{2}\|(L-I)^{-2}u_t\|^2 - 2/3((L-I)^{-2}u,(L-I)^{-1}u) \le 2/3\,E_2(t) .$$

Since the left side of (9.142) is just $1/2$ of the t-derivative of $E_2(t)$, it follows that (9.142) can be rewritten as

$$1/2\frac{d}{dt}E_2(t) \le 2/3\,E_2(t) .$$

Integrating this inequality we obtain (9.141); this completes the proof of Lemma 9.16. ∎

It follows from standard elliptic estimates that $(L-I)^{-2}\delta$ is a continuous function; here δ denotes the Dirac delta function $\delta(x-x', y-y')$. Consequently $((L-I)^{-2}\delta,(L-I)^{-1}\delta)$ is finite for every $\{x', y'\}$. This shows that the initial data (9.136) of R are bounded in the E_2-norm, uniformly for all x' and all y' in a bounded set. Hence using Lemma 9.16 we conclude:

LEMMA 9.17. *The* E_2*-norm of* $R(t)$ *is bounded, uniformly for all* y' *and* t *in a bounded set.*

We need one more basic result from the theory of elliptic partial differential equations.

LEMMA 9.18. *Let* h *be a distribution such that* $(L-I)^N h$ *is locally* L_2 *for all* N. *Then* h *is* C^∞.

We turn now to R smoothed by ϕ; we denote this function by M:

$$M = \int \phi(t)\, R(t)\, dt \ . \tag{9.144}$$

LEMMA 9.19. a) M *is a function of* $x-x'$, y *and* y', *homogeneous of degree zero.*

b) $M = 0$ *if the non-Euclidean distance of* $\{x, y\}$ *from* $\{x', y'\}$ *exceeds* T, *the length of support of* ϕ.

c) M *is* C^∞ *in all of its arguments.*

PROOF. Parts (a) and (b) are corollaries of similar properties of R, listed in Lemma 9.15. To prove (c) we use the fact that R satisfies (9.135), the vector form of the wave equation:

$$R_t = AR \ .$$

Using this relation repeatedly, we get

$$(A^2 - I)^N R = (\partial_t^2 - I)^N R \ .$$

Using this in (9.144) and integrating by parts we get

$$(A^2 - I)^N M = \int [(A^2 - I)^N R]\phi \, dt = \int [(\partial_t^2 - I)^N R]\phi \, dt$$

$$= \int R(\partial_t^2 - I)^N \phi \, dt \ .$$

Since ϕ is assumed to be C^∞, the right member of these equalities is a weighted average of values of $R(t)$, which have, according to Lemma 9.17, a finite E_2-norm. This proves that $(A^2 - I)^N M$ also has a finite E_2-norm. Denoting the components of either of the columns of M by M_1 and M_2 and recalling the action of A, we see that

$$(A^2 - I)^N M = \{(L-I)^N M_1, (L-I)^N M_2\} \ .$$

By (9.140) the E_2-norm of this is

$$\|(L-I)^{N-2} M_2\|^2 - ((L-I)^{N-2} M_1, (L-I)^{N-1} M_1) \ .$$

The boundedness of this quantity for all N shows that M_1 and M_2 belong to the domain of all powers of $L - I$. According to Lemma 9.18 this guarantees that M_1 and M_2 are C^∞ functions of x and y.

To show that M is a C^∞ function of x, y and y' we apply the same argument to $\partial_{y'} R = R'$. This is also a solution of the wave equation (9.135); its initial value is the y' derivative of the initial value of R given by (9.136); that is, it is the derivative of the δ-function.

$$\delta'(x-x', y-y') \begin{pmatrix} 1 & 0 \\ 0 & 1 \end{pmatrix} .$$

This initial value also has a finite E_2-norm and we can proceed as before. However to treat higher order y' derivatives of R, we must introduce higher E-norms:

$$E_n(f) = \|(L-I)^{-n} f_2\|^2 - ((L-I)^{-n} f_1, (L-I)^{-n+1} f_1) \ .$$

The analogue of Lemma 9.16 holds for the E_n-norms and can be proved in an analogous manner.

Proceeding in this fashion, using ever higher E-norms, we can complete the proof of Lemma 9.19. ■

We turn now to the operator

$$\int \phi(t)\, U(t)\, dt \; .$$

The kernel of this operator is

$$M_F = \int \phi(t)\, R_F(t)\, dt \; ,$$

where R_F is the kernel of $U(t)$. R_F is related to R by formula (9.137). Multiplying this with $\phi(t)$ and integrating we get

(9.145) $$M_F(x, y, x', y') = \sum_{\gamma \in \Gamma} M(x, y, \gamma\{x', y'\}) \; .$$

Now according to part (b) of Lemma 9.19,

$$M(x, y, x'', y'') = 0$$

when the distance of $\{x, y\}$ to $\{x'', y''\}$ exceeds T. Since Γ is a discrete subgroup, this will be the case with $\{x'', y''\} = \gamma\{x', y'\}$ for all but a finite number of γ. This shows that the summation on the right in (9.145) is *locally finite*. Since we have shown in part (c) of Lemma 9.19 that M is C^∞, it follows that so is M_F.

According to formula (9.137), the operator K is obtained from $\int \phi U\, dt$ by applying the projection $P_{\mathfrak{D}}$ on the left and right; the action of $P_{\mathfrak{D}}$ is described by equation (9.71). Therefore the kernel of K is obtained from the kernel M_F by composing it with $P_{\mathfrak{D}}$. This amounts essentially to removing the zero Fourier coefficient in x and x' for y and $y' > a$. Clearly, the resulting kernel is still C^∞, except for discontinuities across $y = a$ and $y' = a$ in the cusp neighborhoods. This completes the proof of part (a) of Theorem 9.14.

We turn next to the proof of part (b) of Theorem 9.14. For this we require

LEMMA 9.20. *Suppose* $y > ae^T$. *Then*

(9.146) a) $M_F(x,y,x',y') = 0$ *unless*

$$ye^{-T} < y' < ye^T .$$

(9.147) b) $$M_F(x,x',y,y') = \sum_n M(x-x'+n,y,y').$$

PROOF. The non-Euclidean distance of $\{x,y\}$ to any point $\{x'',y''\}$ with $y'' < ye^{-T}$ or $y'' > ye^T$ is greater than T. Now the parabolic transformations γ pertaining to the cusp in question map the cusp neighborhood $y > a$, $-1/2 < x \le 1/2$ into parallel strips; all other transformations in Γ map points of F below the line $y = a$. It follows from this that if $\{x'y'\}$ is any point with $y' < ye^{-T}$ and $y > ae^T$, then for any γ in Γ, $\{x'',y''\} = \gamma\{x',y'\}$ will satisfy

$$y'' < ye^{-T} .$$

Therefore the non-Euclidean distance of $\{x'',y''\}$ to $\{x,y\}$ exceeds T. According to Lemma 9.19 part (b), this implies that $M(x,y,x'',y'') = 0$. This shows that all terms in (9.145) are zero and proves half of part (a). The other half, pertaining to $y' > ye^T$, goes the same way.

The argument presented above shows that if y' satisfies condition (9.146), then the only nonzero terms in (9.145) come from the parabolic γ which are translations in the x direction by integer amounts. This proves part (b) and with it all of Lemma 9.20. ■

REMARK. The formula (9.147) shows that M_F too is a function of $x-x'$ when $y > ae^T$.

Denote by T^a the operation of removing the zero Fourier coefficient above $y = a$ in each cusp neighborhood. Define K_0 by

$$K_0 = T^a \int \phi(t)\, U(t)\, dt\; T^a\ . \tag{9.148}$$

T^a differs from $P_{\mathfrak{D}}$, defined by (9.71), in the omission of the term $(y/a)^{1/2} f_{1j}^{(0)}(a)$ in the first component. Clearly the difference

$$P_{\mathfrak{D}} - T^a$$

has a finite dimensional range. Comparing K defined by (9.133) and K_0 defined by (9.148), we conclude that $K-K_0$ is a finite dimensional operator.

We have shown that in each cusp the function M_F depends on $x-x'$. This makes it easy to calculate the kernel of K_0 for $y > ae^T$ in each cusp neighborhood as

$$K_0(x, y, x', y') = M_F - \int_{-1/2}^{1/2} M_F\, dx\ .$$

Substituting the sum (9.147) for M_F we get

$$K_0(x, y, x', y') = \sum_{-\infty}^{\infty} M(x-x'+n, y, y') - \int_{-\infty}^{\infty} M(x, y, y')\, dx\ . \tag{9.148$'$}$$

According to part (a) of Lemma 9.19, M is homogeneous of degree zero. Therefore we can rewrite (9.148)′ as

$$K_0 = \sum_{-\infty}^{\infty} M\left(\frac{x-x'+n}{y}, 1, \frac{y'}{y}\right) - y \int_{-\infty}^{\infty} M\left(v, 1, \frac{y'}{y}\right) dv\ . \tag{9.149}$$

Here we have used $v = x/y$ as the new variable of integration.

We now appeal to Lemma 6.8, according to which

$$(9.150)\qquad \left| b \sum_{-\infty}^{\infty} m(bn) - \int_{-\infty}^{\infty} m(v)\,dv \right| \le \text{const. } b^N \int_{-\infty}^{\infty} |\partial_v^N m(v)|\,dv \ .$$

We apply this inequality to

$$m(v) = M\left(\frac{x-x'}{y} + v,\ 1,\ \frac{y'}{y}\right) \quad \text{and} \quad b = 1/y \ .$$

In view of (9.149), the inequality (9.150) can be written in this case as

$$(9.151)\qquad |K_0(x,y,x',y')| \le \frac{\text{const.}}{y^{N-1}} \int \left|\partial_v^N M\left(v,\ 1,\ \frac{y'}{y}\right)\right| dv \ .$$

According to Lemma 9.20, $K_0 = 0$ unless

$$e^{-T} < y'/y < e^{T} \ .$$

Hence we conclude from (9.151) that when $y > a e^T$

$$(9.152)\qquad |K_0(x,y,x',y')| \le \frac{\text{const.}}{(y y')^{\frac{N-1}{2}}} \ .$$

In an entirely analogous fashion we can obtain similar estimates for all derivatives of K_0. Since N in (9.152) is arbitrary, this shows that K_0 belongs to the Schwartz class in the cusp neighborhoods and so completes the proof of Theorem 9.14. ■

REFERENCES

[1] J. Arthur, *The Selberg trace formula for groups of F-rank one*, Annals of Math., vol. 100 (1974), 326-385.

[2] R. Courant and D. Hilbert, *Methods of mathematical physics*, vol. 1, Interscience Publ., New York, 1953.

[3] M. Duflo and J. P. Labesse, *Sur la formule des traces de Selberg*, Ann. scient. Ecole Normale Superieure, vol. 106 (1971), 193-284.

[4] J. Elstrodt, *Die Resolvente zum Eigenwertproblem der automorphen Formen in der hyperbolischen Ebene*, Teil I, Math. Ann., vol. 203 (1973), 295-330; Teil II, Math. Zeitschr., vol. 132 (1973), 99-134; Teil III, Math. Ann., vol. 208 (1974), 99-132.

[5] L. D. Faddeev, *Expansion in eigenfunctions of the Laplace operator in the fundamental domain of a discrete group on the Lobacevskii plane*, Trudy Moscov, Mat. Obsc., vol. 17 (1967), 323-350.

[6] L. D. Faddeev and B. S. Pavlov, *Scattering theory and automorphic functions*, Seminar of Steklov Mathematical Institute of Leningrad, vol. 27 (1972), 161-193.

[7] L. D. Faddeev, V. L. Kalinin and A. B. Venkov, *Nonarithmetic deduction of Selberg's trace formula*, Seminar of the Steklov Mathematical Institute of Leningrad, vol. 37 (1973), 5-42.

[8] I. M. Gelfand, *Automorphic functions and the theory of representations*, Proc. International Congress of Mathematicians, Stockholm (1962), 74-85.

[9] R. Godement, *The decomposition of* $L^2(G/\Gamma)$ *for* $\Gamma = SL(2, Z)$, Proc. of Symposia in Pure Math., Amer. Math. Soc., vol. 9 (1966), 211-224.

[10] ————, *The spectral decomposition of cusp-forms*, Proc. of Symposia in Pure Math., Amer. Math. Soc., vol. 9 (1966), 225-234.

[11] I. C. Gohberg and M. G. Krein, *Introduction to the theory of linear non-selfadjoint operators*, Transl. Math. Monogr., vol. 18, Amer. Math. Soc., Providence, 1969.

[12] T. Kubota, *Elementary theory of Eisenstein series*, John Wiley and Sons, New York, 1973.

[13] R. P. Langlands, *Eisenstein series*, Proc. of Symposia in Pure Math., Amer. Math. Soc., vol. 9 (1966), 235-252.

[14] P. D. Lax and R. S. Phillips, *Scattering theory*, Academic Press, New York, 1967.

[15] P. D. Lax and R. S. Phillips, *The acoustic equation with an indefinite energy form and the Schrödinger equation*, Jr. of Functional Analysis, vol. 1 (1967), 37-83.

[16] ——————, *Scattering theory for dissipative hyperbolic systems*, Jr. of Functional Analysis, vol. 14 (1973), 172-235.

[17] J. Lehner, *Discontinuous groups and automorphic functions*, Amer. Math. Soc. Mathematical Surveys, No. 8, 1964

[18] H. Maass, *Über eine neue Art von nichtanalytischen automorphen Functionen und die Bestimmung Dirichletscher Reihen durch Functionalgleichungen*, Math. Ann., vol. 121 (1949), 141-183.

[19] W. Roelcke, *Über die Wellengleichung bei Grenzkreisgruppen erster Art*, S. B. Heidelberger Akad. Wiss. Math.-Nat. Kl. (1956), 159-267.

[20] ——————, *Analytische Fortsetzung der Eisensteinreichen zu den parabolischen Spitzen von Grenzkreisgruppen erster Art*, Math. Ann., vol. 132 (1956), 121-129.

[21] A. Selberg, *Harmonic analysis and discontinuous groups in weakly symmetric Riemannian spaces with applications to Dirichlet series*, Jr. Indian Math. Soc., vol. 20 (1956), 47-87.

[22] ——————, *Discontinuous groups and harmonic analysis*, Proc. Internat. Congress of Mathematicians, Stockholm (1962), 177-189.

[23] S. Tanaka, *Selberg's trace formula and spectrum*, Osaka Journal of Math., vol. 3 (1966), 205-216.

[24] E. C. Titchmarsh, *Theory of the Riemann Zeta-function*, Clarendon Press, Oxford, 1951.

INDEX

LIBRARY OF CONGRESS CATALOGING IN PUBLICATION DATA

Lax, Peter D

Scattering theory for automorphic functions.

(Annals of mathematics studies; no. 87)

Bibliography: p.

Includes index.

1. Functions, Automorphic. 2. Scattering (Mathematics)
I. Phillips, Ralph Saul, 1913–joint author. II. Title. III. Series.

QA353.A9L39 1976 515'.9 76-3028

ISBN 0-691-08184-0 pbk.

ISBN 0-691-08179-4

www.ingramcontent.com/pod-product-compliance
Ingram Content Group UK Ltd.
Pitfield, Milton Keynes, MK11 3LW, UK
UKHW031027161224
452563UK00001BA/81